Clinical Topics in Hearing
Aid Research

Clinical Topics in Hearing Aid Research

Jason A. Galster, Ph.D.
Katherine E. Stevens, Ph.D.

ISBN 978-1-300-87875-9

Authors
Name: Jason A. Galster
Address: 6600 Washington Avenue South, Eden Prairie, MN 55344
Telephone number: (952) 947-4931
Fax number: (952) 947-4969
E-mail address: jason_galster@starkey.com

Name: Katherine E. Stevens
Address: 462 Half Day Road, Buffalo Grove, IL 60089
Telephone number: (847) 276-2735
Fax number: (847) 276-2733
E-mail address: longgroveaudiology@yahoo.com

Table of Contents

Introduction

Several years ago, Dr. Stevens and I sat down to discuss a collaborative writing project. Although we both have a background in research, Dr. Stevens' day-to-day routine involves the successful ownership of a private audiology practice that focuses on the dispensing of hearing aids; mine concerns the research of hearing aid technology and outcomes. I had a project in mind, but it needed a healthy dose of clinical insight.

This project, and the resulting blog and book, were born from our observation that the average clinician faces many challenges if they wish to follow the hearing aid research literature. Most clinicians spend their workday seeing patients, managing reimbursement, completing documentation, or attending to the myriad needs of running a successful business. With these demands alone, one would be challenged to stay abreast of relevant scientific publications, particularly when membership or subscription is required to access many of our journals.

Even if one were to gain access to all of our hearing-related scientific publications, the task of sifting through this material for clinical insights is a daunting one. For this reason, Dr. Stevens and I set out to review and summarize research articles that offer valuable clinical insight into selecting and fitting hearing aids. Drawing from past and current literature, we created a collection of article reviews that summarize the authors' findings in an accessible manner, describing experimental design and outcomes, and closing each review with commentary on the clinical relevance of their findings.

This book is a topic-driven review of research; some topics consist of three or four consecutive reviews; others are covered with a single review. Our recommendation is that you begin by scanning the Table of Contents for topics that you find interesting. We sincerely hope that you find this book to be a valuable resource with information that is easily applied to your clinical practice.

Physical Characteristics of Hearing Aids

A Comparison of Receiver-in-Canal (RIC) and Receiver-in-the-Aid (RITA) Hearing Aids

Alworth, L.N., Plyler, P.N., Rebert, M.N., & Johstone, P.M. (2010). The effects of receiver placement on probe microphone, performance, and subjective measures with open canal hearing instruments. *Journal of the American Academy of Audiology*, 21, 249–266.

Open-fit behind-the-ear hearing instruments are favored by audiologists and patients alike because of their small size and discreet appearance, as well as their ability to minimize occlusion. The performance of open-fit instruments with the receiver in the aid (RITA) and receiver in canal (RIC) has been compared to unaided conditions and to traditional, custom-molded instruments. However, few studies have examined the effect of receiver location on performance by comparing RITA and RIC instruments to each other. In the current paper, Alworth and her associates (2010) were interested in the effect of receiver location on:

- occlusion;
- maximum gain before feedback;
- speech perception in quiet and noise;
- subjective performance and listener preferences.

Theoretically, RIC instruments should outperform RITA instruments for a number of reasons. Delivery of sound through the thin tube on a RITA instrument can cause peaks in the frequency response, resulting in upward spread of masking (Hoen & Fabry, 2007). Such masking effects are of particular concern for typical open-fit hearing aid users with high-frequency hearing loss. RIC instruments are also capable of a broader bandwidth than RITA aids

(Kuk & Baekgaard, 2008) and may present lowered feedback risk because of the distance between the microphone and receiver (Ross & Cirmo, 1980), and increased maximum gain before feedback (Hallenbeck & Groth, 2008; Hoen & Fabry, 2007).

The authors recruited twenty-five subjects with mild-to-moderate, high-frequency, sensorineural hearing loss to participate in the study. Fifteen had no prior experience with open-canal hearing instruments, whereas ten had some prior experience. Each subject was fitted bilaterally with RIC and RITA instruments with identical signal-processing characteristics, programmed to match NAL-NAL1 targets. Directional microphones and digital-noise-reduction features were deactivated. Subjects used one instrument type (RIC or RITA) for six weeks before testing and then wore the other type for six weeks before being tested again. The instrument style was counterbalanced among the subjects.

Probe microphone measures were conducted to evaluate occlusion and maximum gain before feedback. Speech perception was evaluated with the Connected Speech Test (CST) (Cox et al., 1987), the Hearing in Noise Test (HINT) (Nilsson, et al., 1994), the High Frequency Word List (HFWL) (Pascoe, 1975), and the Acceptable Noise Level (ANL) test (Nabelek et al., 2004). Subjective responses were evaluated with the Abbreviated Profile of Hearing Aid Benefit (APHAB) (Cox & Alexander, 1995), overall listener preferences for quiet and noise, and satisfaction ratings for five criteria: sound quality, appearance, retention and comfort, speech clarity, and ease of use and care.

Real-Ear Occluded Response (REOR) measurements showed minimal occlusion for both types of instruments in this study. Although there was more occlusion overall for RIC instruments, the difference between RIC and RITA hearing instruments was not significant. Overall maximum gain before feedback did not differ between RIC and RITA instruments. However, when analyzed by frequency, the authors found significantly greater maximum gain in the 4000–6000Hz range for RIC hearing instruments.

On the four speech tests, there were no significant differences between RITA versus RIC instruments. Furthermore, there were no significant improvements for aided listening over unaided, except for experienced users with RIC instruments on the Connected Speech Test (CST). It appears that amplification did not significantly improve scores in quiet conditions, for either instrument type, because of ceiling effects. The high unaided speech scores indicated that the subjects in this study, because of their audiometric configurations, already had broad enough access to high-frequency speech cues, even in the unaided conditions. Aided performance in noise was significantly poorer than unaided on the HINT test, but no other significant differences were found for aided versus unaided conditions. This finding was in agreement with previous studies that also found degraded HINT scores for aided versus unaided conditions (Klemp & Dhar, 2008; Valente & Mispagel, 2008).

APHAB responses indicated better aided performance for both instrument types than for unaided conditions on all APHAB categories except aversiveness, in which aided performance was worse than unaided. There were no significant differences between RIC and RITA instruments. However, satisfaction ratings were significantly higher for RIC hearing instruments. New users reported more satisfaction with the appearance of RIC instruments; experienced users indicated more satisfaction with appearance, retention, comfort, and speech clarity. Overall listener preferences were similar, with 80% of experienced users and 74% of new users preferring RIC instruments over RITA instruments.

The findings of Alworth and colleagues (2010) are useful for clinicians and their open-fit hearing aid candidates. Because they provided significantly more high-frequency gain before feedback than RITA instruments, RIC instruments may be more appropriate for patients with significant high-frequency hearing loss. Indeed, this result may suggest that RIC instruments should be the preferred recommendation for open-fit candidates. The results of this study also underscore the importance of using subjective measures with hearing aid patients. Objective speech discrimination testing did not yield significant performance differences between RIC and RITA instruments, but participants showed significant preference for RIC instruments.

Further information is needed to compare performance in noise with RIC and RITA instruments. In this study and others, some objective scores and subjective ratings were poorer for aided conditions than unaided conditions. It is important to note that in the current study, all noise and speech was presented at a 0° azimuth angle, with directional microphones disabled. In real-life environments, it is likely that users would have directional microphones and would participate in conversations with various noise sources surrounding them. Previous work has shown significant improvements with directionality in open-fit instruments (Valente & Mispagel, 2008; Klemp & Dhar, 2008). Future work comparing directional RIC and RITA instruments, in a variety of listening environments, would be helpful for clinical decision making.

Although the performance effects and preference ratings reported here support recommendation of RIC instruments, clinicians should still consider other factors when discussing options with individual patients. For instance, small ear canals may preclude the use of RIC instruments because of retention, comfort, or occlusion concerns. Patients with excessive cerumen may prefer RITA instruments because of easier maintenance and care, or those with cosmetic concerns may prefer the smaller size of RIC instruments. Every patient's individual characteristics and concerns must be considered, but the potential benefits of RIC instruments warrant further examination and may indicate that this receiver configuration should be recommended over slim-tube fittings.

References

Alworth, L.N., Plyler, P.N., Rebert, M.N., & Johstone, P.M. (2010). The effects of receiver placement on probe microphone, performance, and subjective measures with open canal hearing instruments. *Journal of the American Academy of Audiology*, 21, 249–266.

Cox, R.M., & Alexander, G.C., (1995). The Abbreviated Profile of Hearing Aid Benefit. *Ear and Hearing*, 16, 176–186.

Cox, R.M., Alexander, G.C., & Gilmore, C. (1987). Development of the Connected Speech Test (CST). *Ear and Hearing*, 8, 119–126.

Hallenbeck, S.A., & Groth, J. (2008). Thin-tube and receiver-in-canal devices: there is positive feedback on both! *Hearing Journal*, 61(1), 28–34.

Hoen, M., & Fabry, D. (2007). Hearing aids with external receivers: can they offer power and cosmetics? *Hearing Journal*, 60(1), 28–34.

Klemp, E.J., & Dhar, S. (2008). Speech perception in noise using directional microphones in open-canal hearing aids. *Journal of the American Academy of Audiology*, 19(7), 571–578.

Kuk, F., & Baekgaard, L. (2008). Hearing aid selection and BTEs: choosing among various "open ear" and "receiver in canal" options. *Hearing Review*, 15(3), 22–36.

Nabelek, A.K., Tampas, J.W., & Burchfield, S.B. (2004). Comparison of speech perception in background noise with acceptance of background noise in aided and unaided conditions. *Journal of Speech and Hearing Research*, 47, 1001–1011.

Nilsson, M., Soli, S. & Sullivan, J. (1994). Development of the Hearing in Noise Test for the measurement of speech reception threshold in quiet and in noise. *Journal of the Acoustical Society of America*, 95, 1085–1099.

Ross M, Cirmo R. (1980). Reducing feedback in a post-auricular hearing aid by implanting the receiver in an earmold. *Volta Review*, 40–44. Cited in Hallenbeck and Groth, 2008.

Pascoe, D. (1975). Frequency responses of hearing aids and their effects on the speech perception of hearing impaired subjects. *Annals of Otology, Rhinology and Laryngology suppl.* 23, 84: #5, part 2.

Valente, M., & Mispagel, K. (2008). Unaided and aided performance with a directional open-fit hearing aid. *International Journal of Audiology*, 47, 329–336.

Comparing Localization Ability with BTE and CIC Hearing Aids

Best, V., Kalluri, S., McLachlan, S., Valentine, S., Edwards, B., & Carlile, S. (2010). A comparison of CIC and BTE hearing aids for three-dimensional localization of speech. *International Journal of Audiology*, 49(1), 723–732.

Localization of external sound sources is achieved in a number of ways. In additional to visual cues, listeners use binaural time and intensity differences to localize sounds on a horizontal plane (Woodworth, 1938). Monaural spectral cues provide additional information about vertical location and help differentiate sound sources that are in front of or behind the listener (Blauert, 1997). There is ample evidence that localization of sound sources may be an important first step in the perception of speech in complex listening environments (Arbogast et al., 2002; Bregman, 1990; Freyman et al., 2001). Several studies have shown that hearing aid users demonstrate poorer aided localization than when unaided (Byrne et al., 1992; Keidser et al., 2006; Noble & Byrne, 1990; Vanden Bogaert et al., 2006). This is thought to be due to disruption of binaural time and intensity cues by bilateral hearing aids. Therefore, monaural localization cues are valuable to hearing aid wearers and may have particularly important implications for their ability to understand speech in noisy situations.

Two factors known to reduce the availability of monaural spectral cues are of particular relevance to hearing aid users: reduced audible bandwidth (Blauert, 1997; Butler, 1986; Middlebrooks, 1992) and sensorineural hearing loss (Byrne & Noble, 1998; Byrne et al., 1992; 1997; Noble et al., 1994; Rakerd et al., 1998). These factors reduce spectral cue localization because of decreased audibility of high frequencies. Sensorineural hearing loss is accompanied by decreased frequency resolution, which can itself impair spectral cue localization (Jin et al., 2002). Additionally, hearing aid users lose pinna-related spectral cues, particularly with behind-the-ear (BTE) models in which the microphone is placed above the pinna. Completely-in-the-canal (CIC) instruments are thought to preserve pinna-related spectral localization cues because of microphone placement at the ear canal entrance.

The purpose current study was to contrast spatial localization abilities in users with CIC and BTE hearing aids and normal hearing listeners. Two measures of localization were analyzed:

- Lateral localization (horizontal localization: left/right with reference to midline)

- Polar localization (encompassing up/down and front/back dimensions)

The authors recruited eleven subjects with mild-to-moderate sensorineural hearing loss and four subjects with normal hearing. Hearing-impaired subjects were fitted with CIC and BTE instruments. All hearing instruments had 1.5 mm vents and both CIC and BTE instruments had bandwidth out to approximately 6800Hz. Directional microphones, noise-reduction processing, and environment classification features were disabled. Hearing aids were programmed to match CAMEQ gain targets (Moore et al., 1999) and fittings were verified with real-ear measurements. Prior to localization testing, additional probe microphone measurements were conducted to determine aided audibility of the speech stimuli to be used in the test sessions.

Hearing-impaired subjects were tested with both CIC and BTE hearing instruments after a period of "accommodation" or acclimatization to each type of instrument. The experiment was therefore conducted in six phases:

1. Localization testing (both hearing aids)
2. Accommodation period (4–6 weeks, hearing aid A)
3. Localization testing (hearing aid A)
4. Accommodation period (4–6 weeks, hearing aid B)
5. Localization testing (hearing aid B)
6. Localization testing (unaided)

Subjects with normal hearing were tested under two conditions. In one condition, the speech was a broadband stimulus (up to

40,000Hz) and in the other it was low-pass filtered at 6800Hz to approximate the bandwidth of the hearing instruments worn by the hearing-impaired subjects.

Listeners were presented with monosyllabic words at an average level of 65dB SPL for hearing-impaired listeners and 55dB SPL for normal hearing listeners. Subjects were asked to "point their nose" toward the perceived location of the speech. Testing was completed in an anechoic chamber and head orientation was monitored with an electromagnetic tracking system.

The results indicated that for lateral localization errors, there was no difference between CICs and BTEs, no significant difference between aided and unaided results, nor was there a significant effect of accommodation. Performance for normal hearing subjects was more accurate than that of the hearing-impaired subjects. There was a great deal of variability among hearing-impaired subjects; those with poorer low-frequency thresholds had increased lateral localization errors. Previous studies have shown that aided lateral localization is usually worse than unaided and the authors surmised that the performance of the subjects in this study could have been related to their relatively good low-frequency hearing thresholds or the availability of airborne sound through the hearing aid vents.

Analysis of polar angle localization errors yielded similar results. There was no significant effect of hearing aid use, hearing aid style, or accommodation. Performance was substantially better for normal hearing subjects, regardless of bandwidth condition, though errors were slightly greater for the limited bandwidth condition. Although vertical localization in particular was expected to be related to the availability of high-frequency cues, no significant correlational was found for unaided individual performance and high-frequency pure-tone thresholds, or aided results and high-frequency aided sensation level.

Performance with CIC instruments yielded significantly fewer front/back reversals than performance with BTEs and results for both hearing aid types showed significant improvement after accommodation periods. Unaided responses were more accurate than either aided condition and subjects with normal hearing did better than hearing-impaired subjects in any condition. The front/back reversal rate was not correlated with high-frequency audiometric thresholds or aided sensation levels, nor was the benefit of CICs over BTEs correlated with high-frequency sensation level. Previous research shows that front/back localization is primarily related to conchal resonance, which occurs around 4000–5000Hz (Hebrank & Wright, 1974). CIC microphone placement should allow for preservation of these cues, whereas BTE configurations would not. Interestingly, unaided performance in the current study was still better than aided, despite the likelihood that cues in the 4000–5000Hz range would have been inaudible for these subjects without their hearing aids.

The results of this study indicate that hearing-impaired listeners are likely to experience some decreased sound localization ability relative to normal hearing listeners, regardless of hearing aid style. The degree to which localization ability is affected may be related to audiometric thresholds, venting, directionality, compressions settings, and other variables. Though lateral and vertical localization was not affected by hearing aid microphone location in this study, CIC instruments afforded better front/back localization than BTE devices. It is possible that new hearing aid technology will allow for enhanced spectral cue availability. For instance, improvements in feedback control allow more stable high-frequency gain and new, deep-fitting CIC instruments may increase the availability of ear canal and pinna-related spectral cues.

The decrease in front/back localization errors following accommodation periods in this study underscores the importance of acclimatization to new hearing aids. Improvement in localization ability over time is not necessarily something that would warrant adjustments to hearing aid settings, but it should be discussed with new hearing aid users with reference to their expectations during the trial period and thereafter.

Though sound source localization is important for speech perception in complex listening environments, it should be noted that the hearing instruments in this study were programmed without directionality. For many hearing aid users, directional microphones will improve the ability to understand primary speech stimuli in front of the listener so binaural and monaural localization cues may be of decreased significance in some circumstances.

References

Arbogast, T.L., Mason, C.R. & Kidd, G. (2002). The effect of spatial separation on informational and energetic masking of speech. *Journal of the Acoustical Society of America*, 112, 2086–2098.

Best, V., Kalluri, S., McLachlan, S., Valentine, S., Edwards, B., & Carlile, S. (2010). A comparison of CIC and BTE hearing aids for three-dimensional localization of speech. *International Journal of Audiology*, 49(10), 723–732.

Blauert, J. (1997). *Spatial Hearing: The Psychophysics of Human Sound Localization.* Cambridge: MIT Press.

Bregman, A. (1990). *Auditory Scene Analysis.* Cambridge: MIT Press.

Butler, R.A. (1986). The bandwidth effect on monaural and binaural localization. *Hearing Research*, 21, 67–73.

Byrne, D., Noble, W. & LePage, B. (1992). Effects of long-term bilateral and unilateral fitting of different hearing aid types on the ability to locate sounds. *Journal of the American Academy of Audiology*, 3, 369–382.

Byrne, D., & Noble, W. (1998). Optimizing sound localization with hearing aids. *Trends in Amplification*, 3, 51–73.

Freyman, R.L., Balakrishnan, U., & Helfer, K.S. (2001). Spatial release from informational masking in speech recognition. *Journal of the Acoustical Society of America*, 109, 2112–2122.

Hebrank, J., & Wright, D. (1974). Spectral cues used in the localization of sound sources on the median plane. *Journal of the Acoustical Society of America*, 56, 1829–1834.

Jin, C., Best, V., Carlile, S., Baer, T., & Moore, B.C.J. (2002). Speech localization. *Proceedings of the Audio Engineering Society 112th Convention.* Munich, Germany.

Keidser, G., Rohrseitz, K., Dillon, H., Hamacher, V., Carter, L. et al. (2006). The effect of multi-channel wide dynamic range compression, noise reduction and the directional microphone on horizontal localization performance in hearing aid wearers. *International Journal of Audiology,* 45, 563–579.

Middlebrooks, J.C. (1992). Narrow band sound localization related external ear acoustics. *Journal of the Acoustical Society of America,* 92, 2607–2624.

Moore, B.C., Glasberg, B.R., & Stone, M.A. (1999). Use of a loudness model for hearing aid fitting: III. A general method for deriving initial fittings for hearing aids with multi-channel compression. *British Journal of Audiology,* 33, 241–258.

Noble, W. & Byrne, D. (1990). A comparison of different binaural hearing aid systems for localization in horizontal and vertical planes. *British Journal of Audiology,* 24, 335–346.

Noble, W., Byrne, D., & LePage, B. (1994). Effects on sound localization of configuration and type of hearing impairment. *Journal of the Acoustical Society of America,* 95, 992–1005.

Noble, W., Byrne, D., & Ter-Host, K. (1997). Auditory localization, detection of spatial separateness and speech hearing in noise by hearing-impaired listeners. *Journal of the Acoustical Society of America,* 102, 2343–2352.

Rakerd, B., VanderVelde, T.J., & Hartmann, W.M. (1998). Sound localization in the median sagittal plane by listeners with presbyacusis. *Journal of the American Academy of Audiology,* 9, 466–479.

Van den Bogaert, T., Klasen, T.J., Moonen, M., Van Deun, L., & Wouters, J. (2006). Horizontal localization with bilateral hearing aids: without is better than with. *Journal of the Acoustical Society of America,* 119, 515–526.

Woodworth, S. (1938). *Experimental Psychology.* New York: Holt, Rinehart and Winston.

Will Placing a Receiver in the Canal Increase Occlusion?

Vasil-Dilaj, K.A., & Cienkowski, K.M. (2010). The influence of receiver size on magnitude of acoustic and perceived measures of occlusion. *American Journal of Audiology*, 20, 61–68.

The occlusion effect, an increase in bone conducted sound when the ear canal is occluded, is a consideration for many hearing aid fittings. The hearing aid shell or earmold restricts the release of low frequencies from the ear canal (Revit, 1992), resulting in an increase in low-frequency sound pressure level at the eardrum, sometimes up to 25dB (Goldstein & Hayes, 1965; Mueller & Bright, 1996; Westermann, 1987). Hearing aid users suffering from occlusion will complain of an "echo" or "hollow" quality to their voices and hearing their own chewing can be particularly annoying. Indeed, perceived occlusion is reported to be a common reason for dissatisfaction with hearing aids (Kochkin, 2000).

Occlusion from a hearing aid shell or earmold is usually managed by increasing vent diameter or decreasing the length of the vent in order to decrease the acoustic mass of the vent (Dillon, 2001; Kiessling et al., 2005). One potential risk of increasing vent diameter is increased risk of feedback, but this problem has been alleviated by improvements in feedback cancellation. Better feedback management has also resulted in more widespread use of open-fit, receiver-in-canal (RIC) instruments which have proven effective in reducing measured and perceived occlusion (Dillon, 2001; Kiessling et al., 2005; Kiessling et al., 2003; Vasil & Cienkowski, 2006).

Though open-fit BTE hearing instruments are designed to be acoustically transparent, some open fittings still result in perceived occlusion. Interestingly, perceived occlusion is not always strongly or even significantly correlated with measured acoustic occlusion (Kampe & Wynne, 1996; Kiessling et al., 2005; Kuk et al., 2005), so it is apparent that other factors do contribute to the perception of occlusion. The size of the receiver and/or eartip, as well as the size of the ear canal, affect the amount of air flow in and out of the ear canal and it seems likely that these factors could affect the amount of acoustic and perceived occlusion.

Thirty adults, seventeen men and thirteen women, participated in the study. All had normal hearing, unremarkable otoscopic examinations, and normal tympanograms. Two measures of ear canal volume were obtained: volume estimates from the tympanometry screener and estimates determined from earmold impressions that were sent to a local hearing aid manufacturer. Participants were fitted binaurally with RIC hearing instruments. Instead of domes used clinically with RIC instruments, flexible receiver sleeves designed specifically for research purposes were used. Use of the special receiver sleeves allowed the researchers to increase the overall circumference of the receiver systematically so that six receiver size conditions could be evaluated: no receiver, receiver only (with a circumference of 0.149 in.), 0.170 in., 0.190 in., 0.210 in., and 0.230 in.

Real-ear unoccluded and occluded measures were obtained with subjects vocalizing the vowel /i/. Subjects monitored the level of their vocalizations via a sound level meter. Real ear occlusion effect (REOE) was determined by subtracting the SPL levels for the unoccluded response from the occluded response (REOR-REUR = REOE). Subjective measures were obtained by asking subjects to rate their perception of occlusion on a five-point scale ranging from "no occlusion" to "complete occlusion." To avoid bias in the occlusion ratings, participants were not allowed to view the hearing aids or receiver sleeves until after testing was completed.

Results indicated that measured acoustic occlusion was very low for all conditions, especially below 500Hz, where it was below 2dB for most of the receiver conditions. For frequencies above 500Hz, REOE increased as receiver size increased. The *no receiver* and *receiver only* conditions had the least amount of measured occlusion and the largest receiver sizes had the most. There was no significant interaction between receiver size and frequency.

Perceived occlusion also increased as receiver size increased, and though it was mild for most participants in most of the conditions, for the largest receiver condition, some participants rated occlusion as severe. Perceived occlusion was not significantly correlated with measured acoustic occlusion for low frequencies, and the two measures were only weakly correlated for frequencies between 700–1500Hz.

There was no significant relationship between either measure of ear canal volume and perceived or acoustic measures of occlusion. However, adequate ear canal volume to accommodate all receiver sizes was an inclusion criterion for the study, so the authors suggest that smaller ear canal volume could still be a factor in perceived or acoustic occlusion and may warrant further study.

The results of the current investigation show that occlusion was minimal for most of the receiver sizes. These findings are in agreement with previous studies of vented hollow molds, completely open IROS shells (Vasil & Cienkowski, 2006), large 2.4mm vents and silicone ear tips (Kiessling et al., 2005). REOEs for the two largest receivers matched results for a hollow mold with 1mm vent (Kuk et al., 2009) and the REOEs for the two smallest receivers matched results for hollow molds with 2mm and 3mm vents (Kuk et al., 2009). The authors also point out that there was minimal insertion loss for all conditions. Insertion loss from closed earmolds can amount to 20dBHL (Sweetow, 1991) and can contribute to a perception of occlusion or poor voice quality. The relative lack of insertion loss is yet another potential advantage of open and RIC fittings.

Perception of occlusion did increase with the size of the receiver, but overall differences were small. This is in agreement with prior research suggesting that reduction of air flow out of the ear canal results in more low-frequency energy in the ear canal (Rcvit, 1992), which can cause an increase in occlusion (Dillon, 2001). The authors point out that although subjects were not able to see the receivers prior to insertion, they were probably aware of the size and weight differences and could have been influenced by the perception of a larger object in the ear as opposed to actual occlusion. This may also be the case for hearing aid users, perhaps particularly so for individuals with smaller or tortuous ear canals.

The occlusion effect can be challenging, especially when anatomical or other constraints result in the use of minimal venting for individuals with good low-frequency hearing. The results reported here suggest that acoustic occlusion with RIC instruments is slight and may not always be related to perceived occlusion. Therefore, a client's perception of "hollow" voice quality, "echoey" sound quality, or a plugged sensation may be the most reliable indication of occlusion and the most important determinant of eartip size or venting characteristics. The administration of an occlusion rating scale or other self-evaluation techniques may also prove helpful in evaluating occlusion and its impact on overall hearing aid satisfaction.

References

Dillon, H. (2001). *Hearing Aids*. New York, NY: Thieme.

Goldstein, D.P., & Hayes, C.S. (1965). The occlusion effect in bone conduction hearing. *Journal of Speech and Hearing Research, 8*, 137–148.

Kampe, S.D., & Wynne, M.K. (1996). The influence of venting on the occlusion effect. *The Hearing Journal,* 49(4), 59–66.

Kiessling, J., Brenner, B., Jespersen, C.T., Groth, J., & Jensen, O.D. (2005). Occlusion effect of earmolds with different venting systems. *Journal of the American Academy of Audiology,* 16, 237–249.

Kiessling. J., Margolf-Hackl, S., Geller, S., & Olsen, S.O. (2003). Researchers report on a field test of a non-occluding hearing instrument. *The Hearing Journal, 56*(9), 36–41.

Kochkin, S. (2000). MarkeTrak V: Why my hearing aids are in the drawer: The consumer's perspective. *The Hearing Journal,* 53(2), 34–42.

Kuk, F.K., Keenan, D., & Lau, C.C. (2005). Vent configurations on subjective and objective occlusion effect. *Journal of the American Academy of Audiology,* 16, 747–762.

Mueller, H.G., & Bright, K.E. (1996). The occlusion effect during probe microphone measurements. *Seminars in Hearing,* 17(1), 21–32.

Revit, L. (1992). Two techniques for dealing with the occlusion effect. *Hearing Instruments,* 43(12), 16–18.

Sweetow, R. W. (1991). The truth behind "non-occluding" earmolds. *Hearing Instruments,* 42(1), 25.

Vasil, K.A., & Cienkowski, K.M. (2006). Subjective and objective measures of the occlusion effect for open-fit hearing aids. *Journal of the Academy of Rehabilitative Audiology,* 39, 69–82.

Vasil-Dilaj, K.A., & Cienkowski, K.M. (2010). The influence of receiver size on magnitude of acoustic and perceived measures of occlusion. *American Journal of Audiology*, 20, 61–68.

Westermann, V.H. (1987). The occlusion effect. *Hearing Instruments*, 38(6), 43.

Do Over-the-Counter Hearing Aids Offer a Quality Solution to Better Hearing?

Callaway, S.L., and Punch, J.L. (2008). An Electroacoustic Analysis of Over-the-Counter Hearing Aids. *American Journal of Audiology*, 17, 14–24.

Hearing-impaired individuals have a variety of amplification options available to them. Audiologists help patients select the most appropriate hearing aids based on their hearing loss, lifestyle and listening needs, manual dexterity, and a number of other factors. Financial constraints are often a consideration as well, so hearing aid manufacturers offer a wide selection of circuit types, including some more basic, economical choices.

In today's economy, consumers seem more concerned than ever about hearing aid cost. Not surprisingly, there has been an increase in the availability of inexpensive, over-the-counter (OTC) hearing devices. The price of an OTC instrument can range from under fifty dollars to several hundred dollars each. While some of these devices might fit the FDA definition of "hearing aids" (USFDA, 2007a; S874.3300), their distribution often does not meet FDA requirements. For instance, the FDA requires a person buying a hearing aid to be examined by a physician to rule out medical contraindications and a medical waiver must be signed if they choose not to obtain medical clearance. Most OTC devices are purchased in a retail store or over the Internet, so the consumer never interacts with an audiologist and may never be asked for proof of medical clearance. Indeed, the authors of the current study found only one OTC manufacturer that asked consumers to sign a medical waiver prior to purchase.

Despite the decidedly non-clinical distribution of most OTC hearing devices, many audiologists have encountered them, either through advertising or via a patient who has purchased one before coming to our office. Because we recognize the importance of proper diagnosis, selection, fitting verification, and follow-up care, most audiologists have significant reservations about the safety and quality of OTC devices.

These concerns appear to be well founded. The current authors found only one OTC manufacturer that required consumers to submit an audiogram or select an audiometric profile when purchasing an instrument. When they contacted customer service for several OTC manufacturers, they found that representatives had very little knowledge of the technical characteristics of their devices and were unwilling or unable to provide instrument specifications. Previous research has shown that some OTC devices over amplify in the low frequencies and that only reverse-slope hearing losses could be suitably fitted (Cheng & McPherson, 2000).

There is a need for more information about the performance of OTC hearing devices so that audiologists can counsel patients about potential benefits and limitations. In the current study, Callaway and Punch (2008) examined electroacoustic characteristics of eleven OTC hearing devices. The selected OTC instruments were categorized into two groups: low-cost devices priced from $10 to $73 and mid-range devices priced from $349 to $495. The low-cost devices were behind-the-ear, receiver-in-canal style or in-the-canal style. The mid-range devices were in-the-ear or in-the-canal styles.

Technical specification sheets were obtained from the manufacturers of the mid-range devices, but the authors found that specifications for low-cost devices were either unreliable or unavailable. Therefore, they purchased all of the low-cost devices and conducted their own electroacoustic measurements (ANSI 1987, 1996). Tests were conducted twice, two months apart, to ensure reliability and validity of the data. Two of the low-range devices were excluded because they were not working at the time of the second round of testing.

The authors compared NAL-R prescribed gain and output targets (Byrne & Dillon, 1986; Dillon, 2001) to actual gain and outputs measured from the OTC hearing devices. In order to be deemed acceptable for a particular audiometric configuration, gain was required to be within +/-12dB of the NAL targets and output was required to be between -5dB and +3dB of the target. The frequency range was required to provide measurable gain between 250 to 6000Hz.

Overwhelmingly, Callaway and Punch (2008) found that OTC devices had more gain in the low frequencies than in the high frequencies. In fact, all of the low-cost devices were classified as "special purpose" devices because of their low-frequency emphasis and as a result had to be tested using lower three-frequency averages. Total harmonic distortion was within tolerance for all but one OTC instrument, but equivalent input noise was often well above ANSI standards. Only two of the devices, both mid-range devices, had acceptable frequency responses from 250–6000Hz. Most frequency responses were peaked rather than smooth, with some peaks as high as 15dB in the range of 1000Hz to 5000Hz.

Gain and output measures yielded variable results across audiometric configurations for low-cost and mid-range OTC instruments. Because so many of the OTC devices over amplified in the low frequencies, gain targets for the mild-sloping hearing loss configuration were not met, but the flat-moderate hearing loss targets were met more easily. The moderate-sloping loss was the poorest fit, especially for low-cost instruments, primarily because they were unable to provide adequate gain for high frequencies.

The authors concluded that the low-cost instruments were inadequate for use by hearing-impaired individuals because of over-amplification in low frequencies, inadequate high-frequency amplification, high input noise, and narrow frequency response. These conclusions are supported by previous research (Killion, 2003). However, the mid-range instruments had gain and output characteristics that were somewhat more like traditional hearing instruments. Therefore, they could potentially be considered an acceptable low-cost solution for consumers who cannot afford traditional hearing aids dispensed by a hearing care professional. Of course, any of these instruments are more likely to help if consumers are asked to submit recent audiograms or choose an audiometric profile before purchase.

The findings in this study corroborate the concerns many audiologists have about the performance of over-the-counter hearing devices, especially low-cost instruments. In addition to the adverse effects of a reverse sloping, narrow frequency response, high output levels, and frequency response peaks are likely to cause many users of OTC instruments to turn their devices down in order to avoid discomfort or feedback. The resulting reduction in gain would thereby fall even farther below required levels. Because these devices cannot be programmed to an individual's prescribed settings, most users would likely be forced to choose between inadequate gain or discomfort and feedback.

The cost of many OTC hearing devices is low enough that consumers only take a small financial risk if they choose to purchase. However, individuals in need of hearing assistance, having been disappointed with the performance of OTC aids, might assume that appropriately prescribed hearing instruments, fitted and verified by an audiologist, would be no better.

An additional concern regarding the use of OTC products is the fact that purchasers do not get a thorough diagnostic evaluation, nor do they receive recommendations from a qualified hearing care professional. Consumers who forgo a complete audiogram prior to purchasing a hearing device are not referred for appropriate consultation with a physician if they have medical contraindications to hearing aid use or symptoms that require further diagnostic study.

More information about OTC hearing devices is needed, as well as stricter regulation to define and classify them and enforce their proper distribution. More rigorous guidelines should be established to ensure their safety and performance. However, it is also incumbent upon audiologists as hearing care professionals to educate patients about the importance of prescriptive fitting and follow-up care and to guide them to make appropriate decisions about their amplification needs. Over-the-counter hearing devices are bound to appeal to cost-conscious hearing-impaired individuals. Audiologists must be familiar with the limitations and potential risks of OTC devices and be prepared to discuss them with patients.

References

American National Standards Institute (1987). *Specification of hearing aid characteristics*. (ANSI S3.22-1987). New York: Author.

American National Standards Institute (1996). *Specification of hearing aid characteristics*. (ANSI S3.22-1996). New York: Author.

Byrne, D., & Dillon, H. (1986). The National Acoustic Laboratories' (NAL) new procedure for selecting the gain and frequency response of a hearing aid. *Ear & Hearing*, 7, 257–265.

Callaway, S.L., & Punch, J.L. (2008). An Electroacoustic Analysis of Over-the-Counter Hearing Aids. *American Journal of Audiology*, 17, 14–24.

Cheng, C.M., & McPherson, B. (2000). Over-the-counter hearing aids: Electroacoustic characteristics and possible target client groups. *Audiology*, 39(2), 110–116.

Dillon, H. (2001). *Hearing Aids*. New York: Thieme.

Killion, M. (2003). *Citizen petition to the Food and Drug Administration, August 7, 2003. Petition #2003P-0362.* Retrieved from www.fda.gov/ohrms/dockets/dailys/03/aud03/081203/03p-0362-cp00001-vol1.pdf.

U.S. Food and Drug Administration, (2007a). *Subpart D - prosthetic devices*. Retrieved May 6, 2007, from www.accessdata.fda.gov/scripts/cdrh/cfdocs/cfcfr/CFRSearch. 21:8.0.1.1.23.4.

A Preferred Speech Stimulus for Testing Hearing Aids.

Holube, I., Fredelake, S., Vlaming, M., & Kollmeier, B. (2010). Development and analysis of an international speech test signal (ISTS). *International Journal of Audiology, 49*, 891–903.

Current hearing aid functional verification measures are described in the standards IEC 60118 (2005) and ANSI S3.22 (2009) and use static signals, including sine wave frequency sweeps and unmodulated noise signals. Test stimuli are presented to the hearing instrument and frequency-specific gain and output is measured in a coupler or ear simulator. Current standardized measurement methods require the instrument to be set at maximum or a reference test setting and adaptive parameters such as noise reduction and feedback management are turned off.

These procedures provide helpful information for quality assurance and determining fitting ranges for specific hearing aid models. However, they have limitations for today's nonlinear, adaptive instruments and do not provide meaningful information about real-life performance in the presence of dynamically changing acoustic environments.

Speech is the most important stimulus encountered by hearing aid users. Today's nonlinear hearing aids with adaptive signal processing may treat speech differently than stationary signals like sine waves or unmodulated noise. Therefore, it seems preferable for standardized test procedures to use complex stimuli that are as close as possible to natural speech. Indeed, there are some hearing aid test protocols that use samples of natural speech or live speech. But natural-speech stimuli will have different spectra, fundamental frequencies, and temporal characteristics depending on the speaker, the source material, and the language. For hearing aid verification measures to be comparable to each other, it is necessary to have standardized stimuli that can be used internationally.

Alternative test stimuli have been proposed based on the long-term average speech spectrum (Byrne et al., 1994) or temporal envelope fluctuations (Fastl, 1987). The International Collegium for Rehabilitative Audiology (ICRA) developed a set of stimuli (Dreschler, 2001) that reflect the long-term average speech spectrum and have speech-like modulations that differ across frequency bands. ICRA stimuli have advantages over modulated noise and sine wave stimuli in that they share some similar characteristics with speech, but they lack speech-like comodulation characteristics (e.g., fundamental frequency). Furthermore, ICRA stimuli are often classified by signal-processing algorithms as "noise" rather than "speech," so they are less than optimal for measuring how hearing aids process speech.

The European Hearing Instrument Manufacturers Association (EHIMA) is developing a new measurement procedure for nonlinear, adaptive hearing instruments and an important part of their initiative is development of a standardized test signal or International Speech Test Signal (ISTS). The development and analysis of the ISTS was described in a paper by Holube et al., (2010).

There were fifteen articulated requirements for the ISTS, based on available test signals and knowledge of natural speech, the most clinically salient of which include:

- The ISTS should resemble normal speech but should be non-intelligible.

- The ISTS should be based on six major languages, representing a wide range of phonological structures and fundamental frequency variations.

- The ISTS should be based on female speech and should deviate from the international long-term average speech spectrum (ILTASS) for females by no more than 1dB.

- The ISTS should have a bandwidth of 100 to 16,000Hz and an overall RMS level of 65dB.

- The dynamic range should be speech-like and comparable to published values for speech (Byrne et al., 1994; Cox et al., 1988).

- The ISTS should contain voiced and voiceless components. Voiced components should have a fundamental frequency characteristic of female speech.

- The ISTS should have short-term spectral variations similar to speech (e.g., formant transitions).

- The ISTS should have modulation characteristics similar to speech (Plomp, 1984).

- The ISTS should contain short pauses similar to natural running speech.

- The ISTS stimulus should have a sixty-second duration, from which other durations can be derived.

- The stimulus should allow for accurate and reproducible measurements regardless of signal duration.

Twenty-one female speakers of six different languages (American English, Arabic, Mandarin, French, German, and Spanish) were recorded while reading a story, the text and translations of which came from the Handbook of the International Phonetic Association (IPA) (1999). One recording from each language was selected based on a number of criteria including voice quality, naturalness, and median fundamental frequency. The recordings were filtered to meet the ILTASS characteristics described by Byrne et al., (1994) and were then split into 500ms segments that roughly corresponded to individual syllables. These syllable-length segments were attached in pseudo-random order to generate sections of 10 or 15 milliseconds. Each of the resulting sections could be combined to generate different durations of the ISTS stimulus and no single language was used more than once in any six-segment section. Speech interval and pause durations were analyzed to ensure that ISTS characteristics would closely resemble natural speech patterns.

For analysis purposes, a sixty-second ISTS stimulus was created by concatenation of ten- and fifteen-second sections. This ISTS stimulus was measured and compared to natural speech and ICRA-5 stimuli based on several criteria:

- Long-term average speech spectrum (LTASS)

- Short-term spectrum

- Fundamental frequency

- Proportion of voiceless segments

- Band-specific modulation spectra

- Comodulation characteristics

- Pause and speech duration

- Dynamic range (spectral power level distribution)

On all of the analysis criteria, the ISTS stimulus resembled natural speech stimuli as well or better than ICRA-5 stimuli. Notable improvements for the ISTS over the ICRA-5 stimulus were its comodulation characteristics and dynamic range of 20–30dB, as well as pauses and combinations of voiced and voiceless segments that more closely resembled the distributions in natural speech. Overall, the ISTS was deemed an appropriate speech-like stimulus proposal for the new standard measurement protocol.

Following the detailed analysis, the ISTS stimulus was used to measure four different hearing instruments, which were programmed to fit a flat, sensorineural hearing loss of 60dBHL. Each instrument was nonlinear with adaptive noise reduction, compression, and feedback-management characteristics. The first-fit algorithms from each manufacturer were used, with all microphones fixed to an omnidirectional mode. Instead of yielding gain and output measurements across frequency for one input level, the results showed percentile-dependent gain (99th, 65th, and 30th) across frequency as referenced to the long-term average speech spectrum. The percentile-dependent gain values provided information about nonlinearity, in that the softer components of speech were represented by the 30th percentile, moderate and loud speech components were represented by the 65th and 99th percentiles, respectively. Relations between these three percentiles represented the differences in gain for soft, moderate, and loud sounds.

The measurement technique described by Holube and colleagues (2010) using the ISTS stimulus offers significant advantages over current measurement protocols with standard sine wave or noise stimuli. First and perhaps most importantly, it allows hearing instruments to be programmed to real-life settings with adaptive signal-processing features active. It measures how hearing aids process a stimulus that very closely resembles natural speech, so clinical verification measures may provide more meaningful information about everyday performance. By showing changes in percentile-gain values across frequency, it also allows compression effects to be directly visible and may be used to evaluate noise-reduction algorithms as well. The authors also note that the acoustic resemblance of ISTS to speech with its lack of linguistic information may have additional applications for diagnostic testing, telecommunications, or communication acoustics.

The ISTS is currently available in some probe microphone equipment and will likely be introduced in most commercially available equipment over the next few years. Its introduction brings a standardized speech stimulus, for the testing of hearing aids, to the clinic. An important component of clinical best practice is the measurement of a hearing aid's response characteristics. This is most easily accomplished through in-situ probe microphone measurement in combination with a speech-test stimulus such as the ISTS.

References

American National Standards Institute (ANSI). ANSI S3.22-2009. *Specification of hearing aid characteristics.* New York: Acoustical Society of America.

Byrne, D., Dillon, H., Tran, K., Arlinger, S., & Wibraham, K. (1994). An international comparison of long term average speech spectra. *Journal of the Acoustical Society of America, 96*(4), 2108–2120.

Cox, R.M., Matesich, J.S., & Moore, J.N. (1988). Distribution of short-term rms levels in conversational speech. *Journal of the Acoustical Society of America, 84*(3), 1100–1104.

Dreschler, W.A., Verschuure, H., Ludvigsen, C., & Westerman, S. (2001). ICRA noises: Artificial noise signals with speech-like spectral and temporal properties for hearing aid assessment. *Audiology, 40*, 148–157.

Fastl, H. (1987). Ein Storgerausch fur die Sprachaudiometrie. *Audiologische Akustik, 26*, 2–13.

Holube, I., Fredelake, S., Vlaming, M., & Kollmeier, B. (2010). Development and analysis of an international speech test signal (ISTS). *International Journal of Audiology, 49*, 891–903.

International Electrotechnical Commission (IEC), IEC 60118-7 (2005). Hearing Aids: Measurement of electroacoustical characteristics. *Bureau of the International Electrotechnical Commission.* Geneva.

International Phonetic Asssociation (IPA). 1999. *Handbook of the International Phonetic Association.* Cambridge University Press.

Plomp, R. (1984). Perception of speech as a modulated signal. In M.P.R. van den Broeche, A. Cohen (Eds.), *Proceedings of the 10th International Congress of Phonetic Sciences.* (29–40). Utrecht, Dordrecht: Foris Publications, 29-40.

Features of Modern Hearing Aids

Reviewing the Benefits of Open-Fit Hearing Aids

Valente, M., & Mispagel, K.M. (2008). Unaided and aided performance with a directional open-fit hearing aid. *International Journal of Audiology*, 47, 329–336.

With the continued popularity of directional-microphone use in open-fit and receiver-in-canal (RIC) hearing aids, there has been increasing interest in evaluating their performance in noisy environments. A number of studies have investigated the performance of directional, open-fit BTEs in laboratory conditions. (Ricketts, 2000a; Ricketts, 2000b; Valente et al., 1995). Some have evaluated directional-microphone performance in real-life or simulated real-life noise environments (Ching et al., 2009). In the current study, the authors compared performance in omnidirectional, directional, and unaided conditions using RIC instruments in R-Space™ (Revitt et al., 2000) recorded restaurant noise. Their goal was to obtain more externally valid results by using real-life noise in a controlled, laboratory setting.

The R-Space™ method involved recordings of real restaurant noise from an eight-microphone, circular array. For the test conditions, these recordings were presented through an eight-speaker, circular array to simulate the conditions in the busy restaurant. One important factor that distinguishes this study from most others is that the subjects listened to speech stimuli in the presence of noise from all directions, including the front. At the time of this study, only a few other studies had tested directional-microphone performance in the presence of multiple noise sources, including frontal (Bentler et al., 2004; Ricketts, 2000a; Ricketts, 2001).

The authors recruited twenty-six adults with no prior hearing aid experience for the study. They were fitted with binaural receiver-in-canal (RIC) instruments. The instruments were programmed without noise-reduction processing and with independent omnidirectional and directional settings. Subjects were counseled on use and care of the instruments, including proper use of omnidirectional and directional programs. They returned for follow-up adjustments one week after their fitting then used their instruments for four weeks before returning for testing. Subjects were given the opportunity to either purchase the hearing aids after the study at a 50% discount or receive a two hundred dollar payment for participation.

Hearing in Noise Test (HINT) (Nilsson et al., 1994) sentence reception thresholds were obtained to evaluate sentence perception in the uncorrelated R-Space noise. The Abbreviated Profile of Hearing Aid Benefit (APHAB) (Cox & Alexander, 1995) was also administered to evaluate perceived benefit from the instruments in the study. Four APHAB subscales were evaluated independently:

- Ease of communication (EC)
- Reverberation (RV)
- Background noise (BN)
- Aversiveness to loud sounds (AV)

The authors found that subjects' performance in the directional condition was significantly better than both omnidirectional and unaided conditions. The omnidirectional condition was not significantly better than unaided; in fact results were slightly worse than those obtained in the unaided condition.

For the APHAB results, the authors found that on the EC, RV, and BV subscales, aided scores were significantly better than unaided scores. Perhaps not surprisingly, the AV score, which evaluates "aversiveness to noise," was worse in the aided conditions. The aided results combined omnidirectional and directional conditions, so it is possible that aversion to noise in omnidirectional conditions was greater than the directional conditions. However, this was not specifically evaluated in the current study.

The authors pointed out that their directional benefit, which on average was 1.7dB, was lower than those found in other studies of open-fit or RIC hearing instruments (Bentler, 2004; Pumford et al., 2000; Ricketts, 2000b; Ricketts, 2001). However, they mention that most of those studies did not use frontal noise sources in their arrays. Frontal noise sources should have obvious detrimental effects on directional-microphone performance, so it is likely that the speaker arrangement in the current study affected the measured directional improvement. At the time of this publication, one other study had been conducted using the R-Space^{TM} restaurant noise (Compton-Conley et al., 2004). They found mean directional benefits of 3.6 to 5.8 dB, but their subjects had normal hearing and the hearing aids they used were not an open-fit design and were very different from the ones in the current study.

Audiologists can gain a number of important insights from Valente and Mispagel's study. First and foremost, directional microphones are likely to provide significant benefits for users of RIC hearing aids. At the time of publication, the authors noted that directional improvement should be studied in order to warrant the extra expense of adding directional microphones to an open-fit hearing aid order. However, most of today's open-fit and RIC instruments already come standard with directional microphones, many of which are automatically adjustable. So there is no need to justify the use of directional microphones on a cost basis, as they usually add nothing to the hearing aid purchase price.

This study provided more evidence for directional benefit in noise, but further work is needed to determine performance differences between directional and omnidirectional microphones in quiet conditions. Dispensing audiologists should always order instruments that have omnidirectional and directional modes, whether manually or automatically adjustable. This helps ensure that the instruments will perform optimally in most situations. Even instruments with automatically adjustable directional microphones often have push buttons that allow provision of additional programs. For example, a manually accessible, directional program, perhaps with more aggressive noise reduction, offers the patient another option for excessively noisy situations.

The current study obtained slightly reduced directional effects compared to other studies that tested subjects in speaker arrays without frontal noise sources. This underscores the importance of counseling patients about proper positioning when using directional settings. In general, patients should understand that they will be better off when they can put as much noise behind them as possible. But, it is also important to ensure that patients have reasonable expectations about directional microphones. They must understand that the directional microphone will help them focus on conversations in front of them, but will not completely remove competing noise behind them. Patients must also understand that omnidirectional settings are likely to offer no improvement in noise and might even be a detriment to speech perception in some noisy environments.

Subjects in Valente and Mispagel's study (2008) were offered the opportunity to purchase their hearing instruments at a 50% discount after the study's completion. Only eight of the twenty-six subjects opted to do so. Of the remaining subjects, three reported that the perceived benefit was not enough to justify the purchase, whereas fifteen subjects did not report any significant perceived benefit. This leads to another important point about patient counseling.

The subjects in this study, like most candidates for open-fit or RIC instruments, had normal low-frequency hearing. Therefore, they may have had less of a perceived need for hearing aids in the first place. It is important for audiologists to discuss realistic expectations and likely hearing aid benefits with patients in detail at the hearing aid selection appointment, before hearing aids are ordered. Patients who are unmotivated or who do not perceive enough need for hearing assistance will ultimately be less likely to perceive significant benefit from their hearing aids. This is particularly true in everyday clinical situations, in which patients are not typically offered a 50% discount and will have to factor financial constraints into their decisions. For most open-fit or RIC candidates, their motivation and perceived handicap will be related to their lifestyle: their social activities, employment situation, hobbies, etc. Because a patient who has a less than satisfying experience with hearing aids may be reluctant to pursue them again in the future, it is critical for the audiologist to help them establish realistic goals early on, before hearing aid options are discussed.

References

Bentler, R., Egge, J., Tubbs, J., Dittberner, A., & Flamme, G. (2004). Quantification of directional benefit across different polar response patterns. *Journal of the American Academy of Audiology*, 15(9), 649–659.

Ching, T.C., O'Brien, A., Dillon, H., Chalupper, J., Hartley, L., Hartley, D., Raicevich, G., & Hain, J. (2009). *Journal of Speech, Language and Hearing Research*, 52, 1241–1254.

Compton-Conley, C., Neuman, A., Killion, M., & Levitt, H. (2004). Performance of directional microphones for hearing aids: real world versus simulation. *Journal of the American Academy of Audiology*, 15, 440–455.

Cox, R.M., & Alexander, G.C. (1995). The abbreviated profile of hearing-aid benefit. *Ear and Hearing*, 16, 176–183.

Nilsson, M., Soli, S., & Sullivan, J. (1994). Development of the hearing in noise test for the measurement of speech reception thresholds in quiet and in noise. *Journal of the Acoustical Society of America*, 95, 1085–1099.

Pumford, J., Seewald, R., Scollie, S., & Jenstad, L. (2000). Speech recognition with in-the-ear and behind-the-ear dual microphone hearing instruments. *Journal of the American Academy of Audiology*, 11, 23–35.

Revit, L., Schulein, R., & Julstrom, S. (2002). Toward accurate assessment of real-world hearing aid benefit. *Hearing Review*, 9, 34-38, 51.

Ricketts, T. (2000a). The impact of head angle on monaural and bilateral performance with directional and omnidirectional hearing aids. *Ear and Hearing*, 21, 318–329.

Ricketts, T. (2000b). Impact of noise source configuration on directional hearing aid benefit and performance. *Ear and Hearing*, 21, 194–205.

Ricketts, T., Lindley, G., & Henry, P. (2001). Impact of compression and hearing aid style on directional hearing aid benefit and performance. *Ear and Hearing*, 22, 348–360.

Valente, M., Fabry, D., & Potts, L. (1995). Recognition of speech in noise with hearing aids using a dual microphone. *Journal of the American Academy of Audiology*, 6, 440–449.

Valente, M., & Mispagel, K.M. (2008). Unaided and aided performance with a directional open-fit hearing aid. *International Journal of Audiology*, 47, 329–336.

Differences Between Directional Benefit in the Lab and Real-World

Cord, M., Surr, R., Walden, B., & Dyrlund, O. (2004). Relationship between laboratory measures of directional advantage and everyday success with directional microphone hearing aids. *Journal of the American Academy of Audiology, 15*, 353–364.

People with hearing loss require a better signal-to-noise ratio (SNR) than individuals with normal hearing (Bronkhorst & Plomp, 1990; Dubno et al., 1984; Gelfand et al., 1988). Among many technological improvements, a directional microphone is arguably the only effective hearing aid feature for improving SNR and subsequently, improving speech understanding in noise. A wide range of studies support the benefit of directionality for speech perception in competing noise (Agnew & Block, 1997; Nilsson et al., 1994; Ricketts and Henry, 2002; Valente, 1995). Directional benefit is defined as the difference in speech-recognition ability between omnidirectional and directional-microphone modes. In laboratory conditions, directional benefit averages around 7–8dB but varies considerably and has ranged from 2–3dB up to 14–16dB (Agnew & Block, 1997; Valente et al., 1995).

An individual's perception of directional benefit varies considerably among hearing aid users. Cord et al. (2002) interviewed individuals who wore hearing aids with switchable directional microphones and 23% reported that they did not use the directional feature. Many respondents said they had initially tried the directional mode but did not notice adequate improvement in their ability to understand speech and therefore stopped using the directional mode. This discrepancy between measured and perceived benefit has prompted exploration of the variables that affect performance with directional hearing aids. Under laboratory conditions, Ricketts and Mueller (2000) examined the effect of audiometric configuration, degree of high-frequency hearing loss, and aided omnidirectional performance on directional benefit, but found no significant interactions among any of these variables.

The current study by Cord and her colleagues (2004) examined the relationship between measured directional advantage in the laboratory and success with directional microphones in everyday life. The authors studied a number of demographic and audiological variables, including audiometric configuration, unaided SRT, hours of daily hearing aid use, and length of experience with current hearing aids, in an effort to determine their value for predicting everyday success with directional microphones.

Twenty hearing-impaired individuals were selected to participate in one of two subject groups. The "successful" group consisted of individuals who reported regular use of omnidirectional and directional-microphone modes. The "unsuccessful" group of individuals reported not using their directional mode and using their omnidirectional mode all the time. Analysis of audiological and demographic information showed that the only significant differences in audiometric threshold between the successful and unsuccessful group were at 6–8 kHz, otherwise the two groups had very similar audiometric configurations, on average. There were no significant differences between the two groups for age, unaided SRT, unaided word recognition scores, hours of daily use, or length of experience with hearing aids.

Subjects were fitted with a variety of styles—some BTE and some custom—but all had manually accessible omnidirectional and directional settings. The Hearing in Noise Test (HINT) (Nilsson et al., 1994) was administered to subjects with their hearing aids in directional and omnidirectional modes. Sentence stimuli were presented in front of the subject and correlated competing noise was presented through three speakers: directly behind the subject and on each side. Following the HINT, participants completed the Listening Situations Survey (LSS), a questionnaire developed specifically for this study. The LSS was designed to assess how likely participants were to encounter disruptive background noise in everyday situations, to determine if unsuccessful and successful directional-microphone users were equally likely to encounter noisy situations in everyday life. The survey consisted of four questions:

1) On average, how often are you in listening situations in which bothersome background noise is present?

2)   How often are you in social situations in which at least three other people are present?

3)   How often are you in meetings (e.g. community, religious, work, classroom, etc.)?

4)   How often are you talking with someone in a restaurant or dining hall setting?

The HINT results suggest average directional benefit of 3.2dB for successful users and 2.1dB for unsuccessful users. Although directional benefit was slightly greater for the successful users, the difference between the groups was not statistically significant. There was a broad range of directional benefit for both groups: from -0.8 to 6.0dB for successful users and from -3.4 to 10.5dB for the unsuccessful users. Interestingly, three of the ten successful users obtained little or no directional benefit, whereas seven of the ten unsuccessful users obtained positive directional benefit.

Analysis of the LSS results showed that successful users of directional microphones were somewhat more likely than unsuccessful users to encounter listening situations with bothersome background noise and to encounter social situations with more than three other people present. However, statistical analysis showed no significant differences between the two groups for any items on the LSS survey, indicating that users who perceived directional benefit and used their directional microphones were not more likely to encounter noisy situations in everyday life.

The authors concluded that directional benefit as measured in the laboratory did not predict success with directional microphones in everyday life. Some participants with positive directional advantage scores were unsuccessful directional-microphone users and conversely, some successful users showed little or no directional advantage. There are a number of potential explanations for their findings. First, despite the LSS results, it is possible that unsuccessful users did not encounter real-life listening situations in which directional microphones would be likely to help. Directional-microphone benefit is dependent on specific characteristics of the listening environment (Cord et al., 2002; Surr et al., 2002; Walden et al., 2004), and is most likely to help when the speech source is in front of and relatively close to the listener, with spatial separation between the speech and noise sources. Individuals who rarely encounter this specific listening situation would have limited opportunity to evaluate directional microphones and may therefore perceive only limited benefit from them.

Unsuccessful directional-microphone patients may have also had unrealistically high expectations about directional benefits. Directionality can be a subtle but effective way of improving speech understanding in noise. Reduction of sound from the back and sides helps the listener focus attention on the speaker and ignore competing noise. Directional benefit is based on the concept of face-to-face communication; if patients expect their hearing aids to reduce all background noise from all angles, they are likely to be disappointed. Similarly, if they expect the aids to completely eliminate background noise, rather than slightly reduce it, they will be unimpressed. It is helpful for hearing aid users, especially those new to directional microphones, to be counseled about realistic expectations as well as proper positioning in noisy environments. If listeners know what to expect and are able to position themselves for maximum directional effect, they are more likely to perceive benefit from their hearing aids in noisy conditions.

To date, it has been difficult to correlate directional benefit under laboratory conditions with perceived directional benefit. It is clear that directionality offers performance benefits in noise, but directional benefit measured in a sound booth does not seem to predict everyday success with directional microphones. There are many factors that are likely to affect real-life performance with directional-microphone hearing aids, including audiometric variables, the frequency response and gain equalization of the directional mode, the venting of the hearing aid, and the contribution of visual cues to speech understanding (Ricketts, 2000a; 2000b). Further investigation is still needed to elucidate the impact of these variables on the everyday experiences of hearing aid users.

As is true for all hearing aid features, directional microphones must be prescribed appropriately and patients should be counseled about realistic expectations and appropriate circumstances in which they are beneficial. Although most modern hearing instruments have the ability to adjust automatically to changing environments, manually accessed directional modes offer patients increased flexibility and may increase use by allowing the individual to make decisions for improving comfort and performance in noisy places. Routine reinforcement of techniques for proper directional-microphone use is encouraged. Patients should be encouraged to experiment with their directional programs to determine where and when they are most helpful. For the patient, proper identification of and positioning in noisy environments is essential step toward meeting their specific listening needs and preferences.

References

Agnew, J., & Block, M. (1997). HINT thresholds for a dual-microphone BTE. *Hearing Review*, 4, 26–30.

Bronkhorst, A., & Plomp, R. (1990). A clinical test for the assessment of binaural speech perception in noise. *Audiology, 29*, 275–285.

Cord, M.T., Surr, R.K., Walden, B.E., & Olson, L. (2002). Performance of directional microphone hearing aids in everyday life. *Journal of the American Academy of Audiology*, 13, 295–307.

Cord, M., Surr, R., Walden, B., & Dyrlund, O. (2004). Relationship between laboratory measures of directional advantage and everyday success with directional microphone hearing aids. *Journal of the American Academy of Audiology*, 15, 353–364.

Dubno, J.R., Dirks, D.D., & Morgan, D.E. (1984). Effects of age and mild hearing loss on speech recognition in noise. *Journal of the Acoustical Society of America*, 76, 87–96.

Gelfand, S.A., Ross, L., & Miller, S. (1988). Sentence reception in noise from one versus two sources: effects of aging and hearing loss. *Journal of the Acoustical Society of America*, 83, 248–256.

Nilsson, M., Soli, S.D., & Sullivan, J.A. (1994). Development of the Hearing in Noise Test for the measurement of speech reception thresholds in quiet and in noise. *Journal of the Acoustical Society of America*, 95, 1085–1099.

Ricketts, T. (2000a). Directivity quantification in hearing aids: fitting and measurement effects. *Ear and Hearing*, 21, 44–58.

Ricketts, T. (2000b). Impact of noise source configuration on directional hearing aid benefit and performance. *Ear and Hearing* 21, 194–205.

Ricketts, T. (2001). Directional hearing aids. *Trends in Amplification,* 5, 139–175.

Ricketts, T. & Henry, P. (2002). Evaluation of an adaptive, directional microphone hearing aid. *International Journal of Audiology,* 41, 100–112.

Ricketts, T. & Henry, P. (2003). Low-frequency gain compensation in directional hearing aids. *American Journal of Audiology,* 11, 1–13.

Ricketts, T., & Mueller, H.G. (2000). Predicting directional hearing aid benefit for individual listeners. *Journal the American Academy of Audiology,* 11, 561–569.

Surr, R.K., Walden, B.E., Cord, M.T., & Olson, L. (2002). Influence of environmental factors on hearing aid microphone preference. *Journal of the American Academy of Audiology,* 13, 308–322.

Valente, M., Fabry, D.A., & Potts, L.G. (1995). Recognition of speech in noise with hearing aids using dual microphones. *Journal of the American Academy of Audiology,* 6, 440–449.

Walden, B.E., Surr, R.K., Cord, M.T., & Dyrlund, O. (2004). Predicting microphone preference in everyday living. *Journal of the American Academy of Audiology,* 15, 365–396.

What Considerations Should Be Made When Fitting Open-Canal Directional Microphone Hearing Aids?

Klemp, E.J., & Dhar, S. (2008). Speech perception in noise using directional microphones in open-canal hearing aids. *Journal of the American Academy of Audiology*, 19, 571–578.

One of the most common complaints of hearing aid users is difficulty understanding speech in noisy places. Improvements in hearing aid noise reduction may have helped alleviate this problem somewhat, but the most effective solution has been the use of directional microphones (Bentler, 2005).

Behind-the-ear (BTE) hearing aids equipped with directional microphones have been available since the early 1970s. By the mid 1990s, the availability of directional microphones in custom hearing aids gave audiologists the opportunity to offer this benefit to patients who preferred in-the-ear (ITE) styles. In recent years, because of their discreet, unobtrusive appearance and comfortable, lightweight fit, open-canal BTEs have become increasingly popular. Most open-canal instruments on the market today have either automatic or manually accessible directional-microphone programs.

Research with traditional ITE and BTE instruments has shown that the larger the air vent diameter, the more low-frequency sounds are attenuated (Lybarger, 1985). Therefore, enlarged venting has long been a successful method of reducing perceived occlusion for hearing aid users. Because the design of open-fit hearing aids allows for significantly increased venting, the inherent low-frequency attenuation makes them an excellent option for patients with normal hearing in the low- to mid-frequency range and reduces the likelihood of occlusion.

However effective increasing vent diameter may be for reducing occlusion, it has also been demonstrated that the advantage from a directional-microphone system is inversely related to vent diameter in traditional hearing aids (Ricketts, 2000). In other words, as the vent diameter increases, the directional effect decreases. The potential reduction in directional benefit, coupled with the possibility of noise entering the open ear canal raises questions about the performance of open-fit hearing aids in noisy situations. The purpose of Klemp and Dhar's study (2008) was to compare directional hearing aid performance in noise to omnidirectional and unaided conditions.

Sixteen adult subjects with sloping high-frequency hearing losses were tested using the Hearing in Noise Test (HINT) (Nilsson et al., 1994). The HINT sentences were presented at 65dB SPL in the presence of three channels of competing speech-weighted noise. Because the authors intended to evaluate hearing aid performance with active noise reduction, the noise was presented with a twelve-second lead-in to allow hearing aid signal processing to become fully activated before the sentences began.

Subjects were fitted with open-fit hearing aids and were evaluated in five counter-balanced conditions:

- Unaided

- Omnidirectional mode, no digital noise reduction (OMNI)

- Omnidirectional mode with digital noise reduction (DNR)

- Directional mode, no digital noise reduction (DIR)

- Directional mode with digital noise reduction (BOTH)

The analysis of HINT thresholds (in terms of performance and benefit) in these conditions yielded a number of interesting findings, including:

1. Directionality alone and combined with digital noise reduction improved thresholds as compared to omnidirectional (by 3.32dB) or unaided (by 2.26 dB) conditions.

2. Digital noise reduction alone did not improve thresholds.

3. Omnidirectional conditions (with or without noise reduction) yielded poorer thresholds than unaided conditions.

Though the authors pointed out that the directional benefit found with open-fit BTEs in this study is smaller compared to previous findings with traditional occluded fittings (Nordrum et al.,

2006), there was still significant improvement with the use of directional microphones over unaided and omnidirectional conditions. However, they did not find significant improvement with the use of digital noise reduction only, and in some DNR-only trials performance was worse than with omnidirectional amplification alone. These findings support the recommendation and use of directional microphones in open-fit hearing aids. Furthermore, they underscore the importance of combining digital noise-reduction processing with directionality rather than relying on noise reduction alone to improve speech perception in noise.

Perhaps the most interesting finding, however, is the decrement in performance that the authors found in the omnidirectional conditions. As audiologists, our goal is to help patients function better in everyday situations, so we obviously want to avoid recommendations that could result in increased difficulty. Most open-fit BTEs available today have either automatic or manually adjustable directional programs, so it is possible for patients to be in omnidirectional modes in noisy places unless they are counseled thoroughly on the appropriate use of their programs.

Typical candidates for open-fit hearing aids have normal hearing in the low- to middle-frequency range. The hearing aids are not providing low-frequency amplification and are therefore not providing a directional advantage in the low-frequency range. High-frequency directionality has been enhanced in recent hearing instruments by reduction in microphone port spacing (Fabry, 2006). This, along with other signal-processing advances to extend high-frequency response are likely to result in even better performance in noise with open-fit BTE instruments.

As is often the case in our profession, a patient's ultimate success and satisfaction with their open-fit hearing aids may depend on adequate counseling on the use of omnidirectional and directional programs. Even patients who prefer to use automatic programs might benefit from having an additional, manually accessible directional program, for use in situations when the automatic program does not adequately reduce competing noise. Either way, patients need to understand directionality and how their hearing aids are likely to respond to noise backgrounds in everyday conditions so that they can position themselves appropriately and adjust their aids to the proper setting.

References

Bentler, R. (2005). Effectiveness of directional microphones and noise reduction schemes in hearing aids: a systematic review of the evidence. *Journal of the American Academy of Audiology*, 16: 473–484.

Fabry, D. (2006). Facts vs. myths: the "skinny" on slim-tube open fittings: separating truth from fiction in open fittings. *Hearing Review*, May. Retrieved from http://www.hearingreview.com /issues/articles/ 2006-05_04.asp

Klemp, E.J., & Dhar, S. (2008). Speech perception in noise using directional microphones in open-canal hearing aids. *Journal of the American Academy of Audiology*, 19, 571–578.

Lybarger, S. (1985). Earmolds. In J. Katz (Ed.), *Handbook of Clinical Audiology*. (3rd ed.) Baltimore: Williams and Wilkins, 885–910.

Nilsson, M., Soli, S.D., & Sullivan, J.A. (1994). Development of the hearing in noise test for the measurement of speech reception thresholds in quiet and in noise. *Journal of the Acoustical Society of America*, 95(2) 1085–1099.

Nordrum, S., Erler, S., Garstecki, D., & Dhar, S. (2006). Comparison of performance on the hearing in noise test using directional microphones and digital noise reduction algorithms. *American Journal of Audiology*, 15, 81–91.

Ricketts, T. (2000). Directivity quantification in hearing aids: fitting and measurement effects. *Ear and Hearing*, 21, 45–58.

The Real-World Benefits of Directional Microphones with Infants and Young Children

Ching, T.Y.C., O'Brien, A., Dillon, H., Chalupper, J., Hartley, L., Hartley, D., Raicevich, & Hain, J. (2009). Directional effects on infants and young children in real life: implications for amplification. *Journal of Speech Language and Hearing Research*, 52, 1241–1254.

The beneficial effect of directional-microphone use on adult speech perception in noisy environments is well known and is based on the fact that conversational speech usually takes place with participants facing each other. Reducing the level of competing sound behind the listener, even slightly, can increase the signal-to-noise ratio (SNR), resulting in improved identification and discrimination of speech sounds. Clinical audiologists are accustomed to counseling patients to maintain face-to-face contact whenever possible to get the most benefit from the directional microphones and to take advantage of visual cues as well.

The potential advantage of directional-microphone use for children is less understood, partly because children may not employ face-to-face communication as regularly as adults do. Several studies have demonstrated the importance of improved SNR for speech reception in children and it is generally accepted that even children with normal hearing require a greater SNR than adults (Crandell & Smaldino, 2004; Johnstone & Litovsky, 2006). This is particularly true for hearing-impaired children, especially those of a young age. We also know that children are able to orient toward sound sources at a very young age (Ashmead et al., 1987; Muir & Field, 1979; Muir et al., 1989), so it follows that directional microphones could potentially improve their speech reception ability in the presence of competing sounds. However, because of concerns about reduced access to non-frontal speech and environmental sounds, audiologists are often reluctant to fit young children and infants with hearing aids equipped with directional settings for fear of detrimental effects on incidental learning.

Ching et al. (2009) investigated head orientation and the opportunity for young children to benefit from directional hearing aid use in everyday environments. Prior research had shown benefits of directionality in laboratory conditions (Bohnert & Brantzen, 2004; Condie et al., 2002; Kuk et al., 1999), but it was unknown how directionality would affect speech reception in more typical, naturalistic situations. The goal of the study was twofold: 1) to determine the potential benefit of directionality on reception of speech in naturalistic listening situations, and 2) to examine potentially detrimental effects of directionality on non-frontal sounds.

The authors recruited eleven children with normal hearing and sixteen children with moderate hearing loss between the ages of 11 months and 6.5 years. The children were fitted with behind-the-ear, wide dynamic range hearing aids with directional microphones. None of them had prior experience with directionality in their personal hearing aids.

Video recordings of the children were obtained, in four scenarios that represent everyday situations. Diary entries from parents and caregivers were collected to identify listening situations that could account for approximately 80% of child's weekly routine. It was hoped that the diary entries could help predict how often the children were likely to be in situations where directionality could be beneficial.

The video recordings of the children in typical listening scenarios were used to evaluate the proportion of time that they were oriented toward primary speech sources. The four scenarios were:

- The child interacting directly with a caregiver in a play situation

- The child NOT interacting directly with adults in the same room

- The child indoors with other children and adults

- The child outdoors with other children and adults

During the recordings, the researchers logged the time during which speech was "present." Speech was deemed "present" whenever a primary talker could be identified, whether or not they were addressing the child directly.

Video analysis revealed that in the one-to-one situation, the children oriented themselves toward the talker almost 60% of the time. In the remaining group scenarios, the children oriented toward the primary talker between 30–50% of the time, even if they were not being directly addressed by the talker. They were least likely to face the talker in the second scenario, in which adults were present but the child was not engaged in play with adults or other children. Interestingly, age and the presence of hearing loss did not affect the proportion of time that the children spent facing the talker.

Examination of the caregivers' diaries revealed that the majority of the children's time was spent on indoor activities, particularly in group situations. Children with normal hearing were slightly more likely to participate in group activities than hearing-impaired children were. Conversely, hearing-impaired children were somewhat more likely than normal-hearing children to participate in one-to-one activities.

Overall, it was determined that directionality had a positive effect on speech reception, because:

- Children oriented themselves toward the primary talker more than 50% of the time.

- Directionality improved SNR for speech in front of the child, especially in group situations.

- Diary entries showed that the children frequently participated in group activities.

It was also determined that directionality is not likely to have detrimental effects on the perception of incidental speech and environmental sounds. The children still oriented themselves to primary speech sources more than 40% of the time, even when talkers were not directly addressing them. Furthermore, the authors pointed out that the changes in SNR were small, which can be enough to have a significant effect on speech reception from the front in the presence of background noise but is less likely to be enough to affect perception on dominant sound sources from the rear. It follows, then, that directional-microphone settings in hearing aids could have benefits for young pediatric hearing aid users by improving the signal-to-noise ratio and therefore the reception of speech information, especially in group situations.

The authors advised that directional hearing aid programs, partly because of inherent decreases in low-frequency gain, might not always be advisable for children, especially in quiet conditions. They recommended the use of directional settings with equalized frequency responses to adjust for the reduction in low-frequency gain and suggested that switchable instruments would be best, to allow for omnidirectional hearing in quiet conditions and directionality in the presence of noise. Because young children and infants are not capable of adjusting hearing aid settings on their own, automatically adjustable instruments were suggested, especially those that can prioritize speech from a dominant talker even from non-frontal directions. Today we have a wide variety of automatically adjustable directional instruments available at a broad range of price points. This, coupled with ongoing improvements in speech enhancement and noise reduction in hearing aid circuitry indicate that clinicians will have even better tools to help hearing-impaired children function in noisy, everyday situations.

The authors underscored the importance of thoroughly counseling caregivers on the effects of directionality in various listening environments. For instance, caregivers should pay attention to the child's head orientation and positioning and should initiate face-to-face communication at close proximity whenever possible, particularly in noisy situations. Clinical audiologists routinely counsel patients on proper positioning and the importance of face-to-face communication to reduce the effects of background noise on speech perception. Because young, hearing-impaired children rely on better signal-to-noise ratios to receive and process speech information in their everyday activities, and because they may not always orient themselves toward primary speech sources, it is particularly important for their caregivers to understand how they can help maximize the benefit of the child's directional-microphone hearing aids.

References

Ashmead, D.H., Clifton, R.K., & Perrin, E.E. (1987). Precision of auditory localization in human infants. *Developmental Psychology*, 23, 641–647.

Bohnert, A., & Brantzen, P. (2004). Experiences when fitting children with a digital directional hearing aid. *Hearing Review*, 11, 50–55.

Ching, T.Y.C., O'Brien, A., Dillon, H., Chalupper, J., Hartley, L., Hartley, D., Raicevich, & Hain, J. (2009). Directional effects on infants and young children in real life: implications for amplification. *Journal of Speech Language and Hearing Research*, 52, 1241–1254.

Condie, R.K., Scollie, S.D., & Checkley, P. (2002). Children's performance: Analog versus digital adaptive dual-microphone instruments. *Hearing Review*, 9, 40–43.

Crandell, C., & Smaldino, J. J. (2004). Classroom acoustics. In R.D. Kent (Ed.), *The MIT encyclopedia of communication disorders* (442–444). Cambridge, MA: The MIT Press.

Johnstone, P.M. & Litovsky, R.Y. (2006). Effect of masker type and age on speech intelligibility and spatial release from masking in children and adults. *The Journal of the Acoustical Society of America*, 120, 2177–2189.

Kuk, F., Kollofski, C., Brown, S., Melum, A., & Rosenthal, A. (1999). Use of a digital hearing aid with directional microphones in school-aged children. *Journal of the American Academy of Audiology*, 10, 535–548.

Muir, D., & Field, J. (1979). Newborn infants orient to sounds. *Child Development*, 50, 431–436.

Muir, D., Clifton, R.K., & Clarkson, M.G. (1989). The development of a human auditory localization response: A U-shaped function. *Canadian Journal of Psychology*, 3, 199–216.

Considerations for Directional Microphone Use in the Classroom

Ricketts, T., Galster, J., & Tharpe, A.M. (2007). Directional benefit in simulated classroom environments. *American Journal of Audiology*, 16, 130–144.

Classroom acoustic environments vary widely and are affected by a number of factors including reverberation and noise from within the classroom and adjacent areas. Signal-to-noise ratio (SNR) is known to affect speech perception for children with normal hearing and those with hearing loss (Crandell, 1993; Finitzo-Hieber & Tillman, 1978). Because listeners with hearing loss typically require more favorable SNRs to achieve the same performance as normal hearing listeners, hearing-impaired students are particularly challenged by high levels of classroom noise.

FM systems are often recommended as a method for improving SNR in the classroom. However, they may not effectively convey voices other than the teacher's, so children may be less able to hear comments or questions from other students. The additional bulk of ear-level FM systems may prompt reluctance to wear the FM system, as the student may perceive this as calling attention to their hearing loss. Because of these and other potential limitations of FM systems, the use of hearing aids with directional microphones is an opportunity to improve SNR for hearing-impaired children.

The benefits of directional microphones for speech perception in the presence of background noise are well known for adults (Bentler, 2005; Ricketts & Dittberner, 2002; Ricketts et al., 2003). Research has shown that children also benefit from directionality in laboratory conditions (Gravel et al., 1999; Hawkins, 1984; Kuk et al., 1999), but more information is needed on the effect of directional-microphone use in classroom environments. The study summarized in this post evaluated directional-microphone use in simulated classroom situations and the subjective reaction to omnidirectional and directional modes by children and parents.

The authors recruited twenty-six hearing-impaired subjects ranging in age from ten to seventeen years participated in the experiment. All but two had prior experience with hearing aids. Subjects were fitted bilaterally with behind-the-ear hearing instruments that were programmed with omnidirectional and directional modes. Digital noise reduction and feedback-suppression features were disabled and all participants were fitted with unvented, vinyl, full-shell earmolds.

This study consisted of three individual experiments. The first investigated directional versus omnidirectional performance in noise in five simulated classroom scenarios:

1) Teacher Front: speech stimuli presented in front of the listener.

2) Teacher Back: speech presented behind the listener.

3) Desk Work: speech presented in front of the listener, the listener's head oriented down toward desk.

4) Discussion: three speech sources at 0 and 50-degree azimuth (left and right), simulating a round table discussion.

5) Bench Seating: speech presented at 90-degree azimuth (left and right).

Speech-recognition performance was evaluated in each of these scenarios using a modified version of the Hearing in Noise Test for Children (HINT-C) (Nilsson et al., 1994). Speech stimuli were initially presented at 65dB SPL for the five test conditions. Noise was presented from four loudspeakers positioned two meters from each corner of the room. For conditions one through three, the noise level was 55dB SPL. For conditions four and five, noise levels were fixed at 65dB SPL.

A second experiment examined the performance of omnidirectional versus directional modes in the presence of multiple talkers. Monosyllabic words from the NU-6 lists (Tillman & Carhart, 1966) were randomly presented at 63dB SPL from speakers positioned 1.5 meters, surrounding the listener at three angles: 0 degrees (in front of listener), 135 degrees (back right) and 225 degrees (back left). Noise was presented at 57dB SPL, which again yielded an SNR of 6dB.

Not surprisingly, the results of the first experiment showed that directional performance was significantly better than omnidirectional performance for Teacher Front, Desk Work, and Discussion conditions, but was significantly worse for the Teacher Back condition. There was no significant difference between omnidirectional and directional modes for the Bench Seating condition. In the Bench Seating condition, however, subjects were not specifically instructed to look at the speaker. If some subjects did look at the speaker and others did not, individual differences between omnidirectional and directional modes may have been obscured on average. Improved performance was generally noted as the distance between speaker and listener decreased. This is consistent with previous studies with adult listeners, which showed increased directional benefit with decreasing distance (Ricketts & Hornsby, 2003, 2007).

The second experiment yielded no significant difference in performance between omnidirectional and directional modes when speech was in front of the listener. When speech was presented behind the listener, omnidirectional mode was significantly better than directional in both the back-right and back-left conditions. The authors surmised that the directional benefit may have been reduced because subjects were told that all of the talkers were important and because two-thirds of the talkers were behind them, they may have been more focused on speech coming from the back.

The current study offers insight into the potential benefit of directional microphones for classroom environments. An FM system remains the primary recommendation for improving signal-to-noise ratio of a teacher's voice, but overhearing other students and multiple talkers can be compromised by FM technology. Additionally, because of social, cosmetic, or financial concerns, FM use may not be feasible for many students. Therefore, directional-hearing instruments will likely continue to be widely recommended for hearing-impaired schoolchildren. This study reported a directional benefit ranging from 2.2 to 3.3 dB, which is consistent with studies of adult listeners (Ricketts, 2001). Therefore, directional-microphone use in classrooms may indeed be beneficial, as long as the teacher or speaker of interest is in front of the listener. However, for round table or small group arrangements, directionality could be detrimental, especially when talkers are behind the listener. The authors point out that many school scenarios involve multiple talkers or speech from the sides and back, so directional-microphone benefit may be limited overall.

The results of these experiments underscore the importance of counseling for school-age hearing aid users, as well as their parents and teachers. It is common practice to recommend preferential seating close to the teacher in the front of the classroom. Improved performance with decreases in distance from the speech source, in this and other studies, shows that this recommendation is particularly important for hearing aid users, whether or not they are in a directional mode. Furthermore, hearing-impaired students should be instructed to face the teacher so they can benefit from directional processing as well as visual cues. This should also be discussed in detail with teachers so that efforts can be made to arrange classroom seating accordingly.

An incidental finding of the first experiment showed that performance for the Desk Work condition was better than the Teacher Front condition, even though the distance between speaker and listener was comparable. In the Desk Work condition, subjects were instructed to work on an assignment on the desk as they listened. Therefore, the listener's head position was pointed slightly downward, which may have resulted in more optimal, horizontal positioning of the microphone ports, increasing directional effect. This finding demonstrates the importance of selecting the proper tubing or wire length, to position the hearing aid near the top of the pinna and align the microphone ports along the intended plane.

Overall, directional processing improved performance for speech sources in front of the listener and reduced performance for speech sources behind the listener. The instruments in this study were full-time omnidirectional or directional instruments, so it is unknown how automatic, adaptive directional instruments would perform under similar conditions. Because of the prevalence of automatic directionality in current hearing instruments, this is a question with important implications for school-age hearing aid users. Perhaps automatic directionality could provide better overall access to speech in many classroom environments, but controlled study is needed before specific recommendations can be made.

References

Bentler, R.A. (2005). Effectiveness of directional microphones and noise reduction schemes in hearing aids: A systematic review of the evidence. *Journal of the American Academy of Audiology*, 16, 473–484.

Crandell, C. (1993). Speech recognition in noise by children with minimal degrees of sensorineural hearing loss. *Ear and Hearing*, 14, 210–216.

Finitzo-Hieber, T., & Tillman, T. (1978). Room acoustics effects on monosyllabic word discrimination ability for normal and hearing-impaired children. *Journal of Speech and Hearing Research*, 21, 440–458.

Gravel, J., Fausel, N., Liskow, C., & Chobot, J. (1999). Children's speech recognition in noise using omnidirectional and dual-microphone hearing aid technology. *Ear and Hearing*, 20, 1–11.

Hawkins, D.B. (1984). Comparisons of speech recognition in noise by mildly-to-moderately hearing-impaired children using hearing aids and FM systems. *Journal of Speech and Hearing Disorders*, 49, 409–418.

Kuk, F.K., Kollofski, C., Brown, S., Melum, A., & Rosenthal, A. (1999). Use of a digital hearing aid with directional microphones in school-aged children. *Journal of the American Academy of Audiology*, 10, 535–548.

Nilsson, M., Soli, S.D., & Sullivan, J. (1994). Development of the Hearing in Noise Test for the measurement of speech reception thresholds in quiet and in noise. *The Journal of the Acoustical Society of America*, 95, 1085–1099.

Ricketts, T., Lindley, G., & Henry, P (2001). Impact of compression and hearing aid style on directional hearing aid benefit and performance. *Ear and Hearing*, 22, 348–361.

Ricketts, T., & Dittberner, A.B. (2002). Directional amplification for improved signal-to-noise ratio: Strategies, measurement and limitations. In M. Valente (Ed.), *Hearing aids: Standards, options and limitations* (2nd ed., 274–346). New York: Thieme Medical.

Ricketts, T., Galster, J., & Tharpe, A.M. (2007). Directional benefit in simulated classroom environments. *American Journal of Audiology*, 16, 130–144.

Ricketts, T., Henry, P., & Gnewikow, D. (2003). Full time directional versus user selectable microphone modes in hearing aids. *Ear and Hearing*, 24, 424–439.

Ricketts, T., & Hornsby, B. (2003). Distance and reverberation effects on directional benefit. *Ear and Hearing*, 24, 472–484.

Ricketts, T., & Hornsby, B. (2007). Estimation of directional benefit in real rooms: A clinically viable method. In R.C. Seewald (Ed.), *Hearing care for adults: Proceedings of the First International Conference* (195–206). Chicago: Phonak.

Tillman, T., & Carhart, R. (1966). *An expanded test for speech discrimination using CNC monosyllables* (Northwestern University Auditory Test No. 6) SAM-TB-66-55. Evanston, IL: Northwestern University Press.

Understanding the Best Listening Configurations for Telephone Use When Wearing Hearing Aids

Picou, E.M., & Ricketts, T.A. (2010). Comparison of wireless and acoustic hearing aid based telephone listening strategies. *Ear and Hearing, 31*(6), 1–12.

Telephone use is an important consideration for hearing aid users. It is often challenging to arrive at the appropriate coupling method to the ear and related hearing aid settings. Many people with hearing loss have difficulty hearing on the telephone and concerns about telephone use may result in reluctance to purchase new hearing aids or to use aids that have already been purchased (Kochkin, 2000). Indeed, in a survey of hearing aid satisfaction, one in five respondents reported dissatisfaction when using the telephone with a hearing aid (Kochkin, 2005).

There are a number of factors that affect a hearing aid user's ability to hear on the phone, including lack of visual cues, reduced bandwidth, background noise, and difficulty coupling the phone to the hearing aid. The lack of visual cues has been addressed recently with videoconferencing applications, but these are not commonly used, especially among older individuals. The reduced bandwidth (approximately 300 to 3,300 Hz) is characteristic of sound transmission over the phone, so there is little an individual can do improve the availability of high-frequency speech cues over the phone. Background noise and coupling issues can be addressed in a number of ways, depending on the individual and the circumstances.

There are two ways a hearing aid can be coupled directly to the telephone: acoustically and with an inductive telecoil or with acoustic settings that focus on the telephone's limited frequency range. A drawback to the acoustic setting is that the hearing aid microphone is active, which may result in feedback (Latzel et al., 2001; Palmer, 2001; Chung, 2004). Despite recent improvements in feedback control, this remains a problem, especially for those with severe hearing loss whose hearing aids require more gain. Additionally, the microphone picks up environmental noise that competes with the telephone signal, decreasing the signal-to-noise ratio. Telecoils can be a solution for feedback and poor signal-to-noise ratios, but they are subject to interference from fluorescent lights, computer equipment, and power lines. Furthermore, it can be difficult to determine the proper positioning of the phone for optimal sound quality, as the telephone receiver must be placed as close to the telecoil as possible (Compton, 1994; Tannahill, 1983; Yanz & Preves, 2003).

A recent option for telephone is through the use of intermediate wireless accessories, these route sound from the phone to the hearing aids via a combination of Bluetooth and a direct-to-hearing aid wireless technology. These devices address the problems with acoustic or telecoil coupling, and have the possibility of providing some additional benefit if the telephone signal is bilaterally routed (Green, 1976; Hall et al., 1984; Moore, 1998; Quaranta and Cervellera, 1974). Many hearing aid manufacturers offer wireless devices, but it is unclear whether their use results in significantly improved speech recognition over the phone. Even with wireless routing of the phone signal, there may still be detrimental effects of background noise, especially for individuals with open-canal hearing aids (Dillon, 1985, 1991).

The purpose of Picou and Ricketts' study (2010) was to examine speech-recognition performance with monaural and binaural wireless phone transmission, as well as a monaural acoustic condition, in the presence of two levels of background noise. They also evaluated performance with occluding versus non-occluding domes.

Twenty individuals with sloping, high-frequency, sensorineural hearing loss participated in the study. Subjects were fitted with binaural, receiver-in-canal hearing instruments with a wireless transmitter accessory. Half of the subjects were tested with open, non-occluding domes and half were tested with closed, occluding domes.

A total of seven hearing aid and telephone configurations were tested in two background noise levels (55dBA SPL and 65dBA SPL). Subjects responded to sentences from the Connected Speech Test (CST), (Cox et al., 1987). Speech stimuli were band pass filtered from 300 to 3400Hz to simulate telephone transmission and presented at

65dB SPL. Competing speech babble was presented through four loudspeakers positioned around the listener at a distance of one meter. All test conditions—hearing aid condition, dome type, noise level—were counterbalanced to avoid effects of learning and fatigue.

This study illuminates some important considerations in telephone use and supports the use of wireless telephone accessories, especially with bilateral routing. The participants performed best with external hearing aid microphones turned off, but the authors acknowledge that for safety and monitoring of environmental sounds, it may be advisable to leave microphones active at an attenuated level. The authors suggest that further investigation is warranted to determine optimal levels of microphone attenuation to allow for successful speech recognition over the phone, while preserving environmental awareness.

Performance with occluding domes was better than open domes for wireless telephone signal routing in noise. Occluding domes reduce the environmental noise entering the ear canal, providing an improvement in signal-to-noise ratio. In the acoustic phone condition, open domes performed better than occluding domes. Subjects tended to position the phone directly over the ear canal, which likely improved signal-to-noise ratio by blocking background noise and isolating the speech transmitted from the phone.

Specific observations were made for participants wearing open-canal hearing aids. Specifically, users with open domes should be instructed to hold the phone directly over the ear canal for optimal speech recognition. Programming adjustments may be necessary to increase availability of low- and mid-frequency speech cues and improve signal-to-noise ratio. Conversely, users with occluding domes should be advised of the potential limitations of direct acoustic coupling to the phone and should be instructed to hold the phone receiver as close to the microphone as possible. Alternatively, patients with occluding domes may be better off using a telecoil, if available, for situations in which they cannot use a wireless device.

Interestingly, no significant improvement in speech recognition resulted from plugging the non-test ear or muting the hearing aid on the non-test ear. This is consistent with previous research on masking level differences for tones (Green, 1976; Moore, 1998) as well as a previous study of speech recognition over the phone, which found no improvement for normal-hearing listeners when the non-phone ear was plugged. This is inconsistent, however, with reported preferences of hearing aid users. Despite the lack of improvement in the current study, the authors acknowledged that muting the hearing aid on the non-phone ear may reduce listening effort, which is therefore preferred by the listener.

For users of wireless accessories, the results of this study clearly indicate that binaural routing is ideal. But for hearing aid users who do not have wireless devices, the optimal hearing aid settings and coupling method may depend on several factors. The extent of venting or openness should be considered when choosing an acoustic phone coupling; individuals with minimal venting may not hear well unless they are able to hold the telephone over the hearing aid microphone, while patients with open fittings may experience more challenges with background noise interference than the more occluded wearer.

Regardless of whether a client uses an intermediate wireless device for binaural telephone streaming, monaural acoustic listening, or telecoil coupling, the attenuation level of the hearing aid microphones is also a consideration. For binaural wireless routing or streaming, it is advisable to keep both hearing aid microphones active but attenuated, to preserve awareness of environmental sounds. For monaural acoustic/telecoil combinations the microphone level on the opposite ear can be attenuated slightly to allow environmental awareness but reduce distraction from surrounding noise. As noted earlier, further study is warranted to determine optimal microphone attenuation levels.

References

Chung, K. (2004). Challenges and recent developments in hearing aids. Part II. Feedback and occlusion effect reduction strategies, laser shell manufacturing processes and other signal processing technologies. *Trends in Amplification, 8*, 125–164.

Compton, C. (1994). Providing effective telecoil performance with in-the-ear hearing instruments. *Hearing Journal, 47*, 23–26.

Cox, R.M., Alexander, G.C., & Gilmore, C.A. (1987). Development of the connected speech test (CST). *Ear and Hearing, 8* (supplement): 119S–126S.

Dillon, H. (1985). Earmolds and high frequency response modification. *Hearing Instruments,* 36, 8–12.

Dillon, H. (1991). Allowing for real ear venting effects when selecting the coupler gain of hearing aids. *Ear and Hearing,* 12(6), 406–416.

Green, D.M. (1976). *An Introduction to Hearing.* Hillsdale, NJ: Lawrence Erlbaum Associates.

Hall, J.W., Tyler, R.S., Fernandes, M.A. (1984). Factors influencing the masking level difference in cochlear hearing-impaired and normal-hearing listeners. *Journal of Speech and Hearing Research,* 27, 145–154.

Kochkin, S. (2000). MarkeTrak V: "Why my hearing aids are in the drawer": The consumers' perspective. *Hearing Journal,* 53, 34–42.

Kochkin, S. (2005). MarkeTrak VII: Customer satisfaction with hearing aids in the digital age. *Hearing Journal,* 58, 30–39.

Latzel, M., Gebhart, T.M. & Kiessling, J. (2001). Benefit of a digital feedback suppression system for acoustical telephone communication. *Scandanavian Audiology Supplementum,* 52, 69–72.

Moore, B.C.J. (1998). *Cochlear Hearing Loss.* London: Whurr Publishers.

Palmer, C.V. (2001). Ring, ring! Is anybody there? Telephone solutions for hearing aid users. *Hearing Journal,* 54, 10.

Picou, E.M., & Ricketts, T.A. (2010) Comparison of wireless and acoustic hearing aid based telephone listening strategies. *Ear and Hearing,* 31(6), 1–12.

Quaranta, A., & Cervellera, G. (1974). Masking level difference in normal and pathological ears. *Audiology,* 13, 428–431.

Tannahill, J.C. (1983). Performance characteristics for hearing aid microphone versus telephone and telephone/telecoil reception modes. *Journal of Speech and Hearing Research,* 26, 195–201.

Yanz, J.L., & Preves, D. (2003). Telecoils: Principles, pitfalls, fixes and the future. *Seminars in Hearing,* 24, 29–41.

The Effect of Digital Noise Reduction on Listening Effort

Sarampalis, A., Kalluri, S., Edwards, B., & Hafter, E. (2009). Objective measures of listening effort: Effects of background noise and noise reduction. *Journal of Speech Language and Hearing Research*, 52, 1230–1240.

In this 2009 study, the authors pursue the sometimes-elusive benefits of digital noise reduction. A review of past literature suggests that digital noise reduction, as implemented in hearing aids, benefits patients through improved sound quality, ease of listening, and a possible perceived improvement in speech understanding. Significant improvements in speech understanding are, however, not a routinely observed benefit of digital noise reduction and some studies have shown significant decreases in speech understanding with active digital noise reduction.

In a 1992 article, authors Hafter and Schlauch suggest that noise reduction may lighten a patient's cognitive load, essentially freeing resources for other tasks. To better understand the proposed effect, imagine driving a car in an unfamiliar area. It's common for drivers to turn their stereo down, or off, when driving in a demanding situation. This is beneficial, not because music affects driving ability, but because the additional auditory input is distracting, effectively increasing the driver's cognitive load. By removing the distraction of the stereo, more cognitive resources are freed and the ability to focus, or pay attention to the complex task of driving, is improved.

In order to better understand how digital noise reduction may affect attention and cognitive load, two experiments were completed. In the first experiment, research participants were asked to repeat the last word of sentences presented in a background of noise. After eight sentences, the listener attempted to repeat as many of the target words as they could. The sentence material contained both high-context and no-context conditions, for example:

High context: *A chimpanzee is an ape.*

No context: *She might have discussed the ape.*

In the second experiment, listeners were asked to judge if a random number between one and eight was even or odd, while at the same time listening to and repeating sentences presented in a background of noise. Both experiments incorporated a dual-task paradigm: the first asked participants to repeat select words presented in noise, while also remembering these words for later recall. The second required participants to repeat an entire sentence, presented in noise, while also completing a complex visual task.

Highlights from experiment one show:

- Performance in all conditions decreased as the signal-to-noise ratio became more difficult.

- Overall performance in the no-context conditions was lower than in the high-context conditions.

- A comparison between performance with and without digital noise reduction showed a significant improvement in recall ability with digital noise reduction.

Highlights from experiment two show:

- Performance in all conditions decreased as the signal-to-noise ratio became more difficult.

- Reaction times increased with decreased signal-to-noise ratio.

- At -6 dB SNR, reaction times were significantly improved with digital noise reduction.

The findings of this study show that the cognitive demands of non-auditory tasks, such as visual and memory tasks, inhibit the ability of a person to understand speech in noise. In other words, secondary tasks make speech understanding more difficult. Additionally, digital-noise-reduction algorithms can reduce cognitive effort under adverse listening conditions. The authors discuss the value of using cognitive measures in hearing aid research and speculate that directional microphones may provide a cognitive benefit as well.

The clinical implications of this study suggest that patients may find benefits of wearing hearing aids that go beyond improved speech audibility. Modern signal processing may provide benefits that are only now being understood. For instance, a patient may report that hearing aids have made listening easier, that their new hearing instruments seem to suppress noise more than the old ones, but routine evaluation of speech understanding may not show significant differences between the two hearing aids.

Hearing aid success and benefit has traditionally been defined with the results of speech testing, or questionnaires. If advanced technology can ease the task of listening, patients may be receiving benefits from their hearing aids that we are not currently prepared to evaluate in the office or clinic. Future work in this area will increase the understanding of the role that cognition plays in the success of the hearing aid wearer.

References

Hafter, E. R., & Schlauch, R. S. (1992). Cognitive factors and selection of auditory listening bands. In A. Dancer, D. Henderson, R. J. Salvi, & R. P. Hammernik (Eds.), *Noise-induced hearing loss* (303–310). Philadelphia: B.C. Decker.

Sarampalis, A., Kalluri, S., Edwards, B., & Hafter, E. (2009). Objective measures of listening effort: Effects of background noise and noise reduction. *Journal of Speech Language and Hearing Research*, 52, 1230–1240.

Effects of Digital Noise Reduction on Children's Speech Understanding

Stelmachowicz, P., Lewis, D., Hoover, B., Nishi, K., McCreery, R., & Woods, W. (2010). Effects of digital noise reduction on speech perception for children with hearing loss. *Ear & Hearing*, 31, 345–355.

Because a great deal of everyday communication takes place in the presence of some level of background noise, hearing aid performance in noise is of interest to researchers, clinicians, and hearing aid users. It is well established that directional microphones can improve signal-to-noise ratio (SNR) for adult hearing aid users as well as children (Gravel et al. 1999; Valente et al. 1995). It is generally accepted that digital noise reduction (DNR) will not improve speech recognition in noise (Bentler et al., 2008; Levitt et al., 1993). Digital noise reduction has, however, resulted in improved overall sound quality judgments and decreased listening effort (Boymans & Dreschler, 2000; Sarampalis et al., 2009; Walden et al., 2000).

In noisy situations, adult listeners use a variety of cues to understand conversational speech, including visual cues, situational cues, and semantic and grammatical context. Young children with limited language skills may not be able to take advantage of this information and may rely more on acoustic cues. Indeed, most studies show that children require better SNRs than adults (Blandy & Lutman, 2005; Jamieson et al., 2004).

For hearing-impaired children, hearing aids are more than a tool for the recognition of speech; they facilitate speech and language acquisition and development. As the authors of the current study pointed out, "Amplification must facilitate the development of early auditory skills, laying the foundation for the extraction of regularities in the speech signal and the development of language." Therefore, improving access to speech is of particular importance for young hearing aid users. Conversely, it is also must be determined that DNR or directional processing is not degrading the speech signal.

The purpose of the present study was to determine the effect of DNR on children's perception of nonsense syllables, words, and sentences in the presence of noise. Sixteen children with mild-to-moderately severe hearing loss participated in the study. Subjects were divided into two groups: five- to seven-year-olds and eight- to ten-year-olds. The authors chose these age groups to evaluate the effect of development on the perception of speech stimuli with varying levels of context. Subjects were fitted with binaural behind-the-ear hearing aids with DNR and amplitude compression; all testing including DNR-on and DNR-off conditions. Directional microphones were not activated. Hearing aids were programmed to DSL 5.0 targets and settings were verified with real-ear measurements.

The children were presented with speech stimuli mixed with speech-shaped noise at SNRs of 0dB, +5dB and +10dB. Three levels of context were represented:

- VCV (vowel-consonant-vowel) nonsense syllables, fifteen consonants combined with /a/

- Monosyllabic words from the Phonetically Balanced Kindergarten List (PBK) (Haskins, 1949)

- Meaningful sentences with three key words each (Bench et al. 1979)

Data analysis revealed that noise reduction did not have a significant positive or negative effect on performance. There was no significant main effect for context, but not surprisingly, post hoc testing revealed that scores for both age groups were higher for sentences than they were for both nonsense syllables and monosyllables. Also not surprisingly, performance improved with increases in SNR for all types of speech stimuli. There was a significant effect of age, with older subjects demonstrating better overall performance than younger subjects. There was no interaction between age and noise reduction, indicating that the use of noise reduction did not affect performance of younger and older subjects differently. There was no interaction between age group and context, indicating that both age groups benefitted from context equivalently.

The authors observed a great deal of variability among subjects, especially the younger group. Though noise reduction did not significantly affect performance overall, the authors found that more than half of the younger subjects demonstrated poorer recognition of words in the DNR-on condition. The most common consonant confusions were: /f/ for /t/, /g/ for /d/, and /b/ for /v/, suggesting that voicing information was perceived correctly but place and manner of articulation were not easily distinguished. This finding is in agreement with previous results reported by Jamieson et al. (1995), who found that DNR processing resulted in either no improvement or a slight decrement in performance and that consonant place of articulation was particularly affected. There are several cues that affect consonant perception and slight decrements in the acoustic representation of a consonant may be offset by the availability of other cues. For example, though /f/ and /t/ may be difficult to discriminate, a participant in face-to-face conversation benefits from visual cues to help identify these consonants. Still, the opportunity exists to further study the effect of noise reduction on consonant perception with adult and pediatric subjects.

Despite the minimal effect of noise reduction on speech recognition, all listeners in Jamieson's study (1995) reported a strong preference for DNR processing when hearing continuous speech in a variety of listening environments. This leads to an important consideration regarding the use of noise-reduction processing in hearing aids for children. Although the current investigation did not address listening preference, previous studies with adults have often shown positive effects of noise-reduction processing on listening effort and sound quality. The current authors suggested that if this were also the case for children, it could improve attentiveness and increase "time on task" in difficult listening situations. This is an interesting hypothesis, since attention and focus is essential for understanding speech in noise and many hearing-impaired children may demonstrate attention deficits.

Audiologists working with pediatric patients should consider noise-reduction settings carefully. Although there were no statistically significant effects of noise reduction on speech perception in this study, decreases in word recognition scores for younger children in the DNR-on condition is a concern and warrants further study. The authors point out that a child's ability to recognize and understand speech requires ongoing, consistent auditory experiences. Previous use of amplification, age of identification, and consistency of hearing aid use may have influenced the results of this study and may affect success with DNR processing in general. The effect of degree of hearing loss should also be considered, as it is possible that individuals with severe hearing losses could be adversely affected by even small decrements in speech information resulting from DNR processing.

Clinically, an important highlight of this study is the fact that individual performance among children is highly variable. Digital noise reduction has the potential to ease listening, but may compromise clarity of speech. And directional microphones may improve access to speech, but also risk compromising speech audibility for off-axis talkers. These considerations suggest that some advanced features should be reserved for older children and specific environments. Among that older population, there may be some inclination to allow manual adjustment of hearing aid settings. However, Ricketts and Galster (2008) correctly point out that children cannot be expected to adjust manual directionality controls reliably. This ultimately results in a fitting rationale that avoids the fitting of some advanced features or allows them to function automatically, with the assumption that they will only be active in the appropriate situations and "do no harm" in regard to speech recognition.

Further study of the perceptual effects of noise reduction and subjective preferences in children is needed. The possibility remains that DNR may offer hearing-impaired children other benefits such as improved attention and comfort in noise, possibly leading to increased satisfaction and compliance from pediatric patients.

References

Bench, J., Kowal, A., & Bamford, J. (1979). The BKB sentence lists for partially-hearing children. *British Journal of Audiology*, 13, 108–112.

Bentler, R., Wu, Y.H., Kettel, J. (2008). Digital noise reduction: Outcomes from laboratory and field studies. *International Journal of Audiology*, 47, 447–460.

Blandy, S. & Lutman, M. (2005). Hearing threshold levels and speech recognition in noise in 7-year-olds. *International Journal of Audiology*, 44, 435–443.

Boymans, M., & Dreschler, W.A. (2000). Field trials using a digital hearing aid with active noise reduction and dual-microphone directionality. *Audiology*, 39, 260–268.

Gravel, J.S., Fausel, N., Liskow, C. (1999). Children's speech recognition in noise using omnidirectional and dual microphone hearing aid technology. *Ear and Hearing*, 20, 1–11.

Haskins, H.A. (1949). A phonetically balanced test of speech discrimination for children. Master's thesis, Northwestern University, Evanston, IL.

Jamieson, D.G., Kranjc, G., Yu, K. (2004). Speech intelligibility of young school-aged children in the presence of real-life classroom noise. *Journal of the American Academy of Audiology*, 15, 508–517.

Levitt, H., Bakke, M., Kates, J. (1993). Signal processing for hearing impairment. *Scandanavian Audiology Supplement*, 38, 7–19.

Ricketts, T.A., & Galster, J. (2008). Head angle and elevation in classroom environments: implications for amplification. *Journal of Speech, Language and Hearing Research*, 15, 516–525.

Sarampalis, A., Kalluri, S., Edwards, B., & Hafter, E. (2009). Objective measures of listening effort: Effects of background noise and noise reduction. *Journal of Speech Language and Hearing Research*, 52, 1230–1240.

Stelmachowicz, P., Lewis, D., Hoover, B., Nishi, K., McCreery, R., & Woods, W. (2010). Effects of digital noise reduction on speech perception for children with hearing loss. *Ear & Hearing*, 31, 345–355.

Valente, M., Fabry, D., Potts, L.G. (1995). Recognition of speech in noise with hearing aids using dual microphones. *Journal of the American Academy of Audiology*, 6, 440–449.

Walden, B.E., Surr, R.K., Cord, M.T. (2000). Comparison of benefits provided by different hearing aid technologies. *Journal of the American Academy of Audiology*, 11, 540–560.

Does Expansion Decrease Low-Level Speech Understanding?

Brennan, M., & Souza, P. (2009). Effects of expansion on consonant recognition and consonant audibility. *Journal of the American Academy of Audiology*, *20*, 119–127.

The primary goal of a hearing aid fitting is to improve audibility and availability of speech sounds while maintaining comfort and loudness tolerance. A linear hearing aid fitting may provide audibility for average speech sounds but may result in discomfort for loud sounds and inaudibility for soft sounds. The use of wide-dynamic range compression (WDRC) has addressed these issues, helping maximize the useful dynamic range of hearing for individuals who require amplification for quiet and moderate sounds, yet have limited tolerance for loud sounds.

One potential issue with WDRC has been the increased audibility of very soft environmental sounds, which may be unwelcome for individuals who have adjusted to long-term hearing loss and are not used to perceiving these sounds. An additional problem for individuals with good residual hearing at some frequencies is that they will hear circuit noise from the hearing aid itself. Both of these issues can be unpleasant for the listener, possibly resulting in rejection or limited use of the hearing aids.

Expansion makes hearing aids quieter at low input levels. This is done in almost all modern hearing aids in order to reduce annoying environmental or circuit noise. There is concern, however, that if too aggressive, expansion can have a detrimental effect on speech intelligibility (Plyler et al., 2005). It has been proposed that reduced speech-recognition ability with expansion is due to reduced audibility of speech cues (Plyler et al., 2007; Walker et al., 1984).

Previous examinations of expansion have measured its effect on audibility of room noise (Plyler et al., 2005) or the long-term average speech spectrum (LTASS) (Zakis & Wise, 2007) but did not directly measure the effect of expansion on audibility of the speech signals. The current authors sought more specific insight into the effect of expansion on speech recognition by studying the relationship between expansion and consonant audibility.

Though there may be other related parameters warranting examination, the primary variables of interest relating to expansion are the ratio and the kneepoint. In theory, a high-expansion kneepoint should have a negative effect on speech recognition, because gain for stimuli below the kneepoint is reduced, resulting in decreased audibility. Speech presented above the expansion kneepoint should be less affected by the expansion.

Therefore, the hypotheses for Brennan and Souza's study (2009) were as follows:

1. A high-expansion kneepoint will significantly reduce consonant-vowel (CV) recognition.

2. A high-expansion kneepoint will significantly reduce CV audibility.

3. The effect of expansion on speech recognition and audibility will be reduced for increased speech input levels.

4. There will be a significant positive correlation between CV recognition and audibility for each condition.

Thirteen hearing-impaired individuals participated in the experiment. Nine were experienced hearing aid users; the remaining four did not use hearing aids. Subjects were fitted monaurally with a multi-channel, digital, behind-the-ear hearing aid. Venting was 3mm for most subjects, but was reduced to 1mm for two subjects and plugged for one subject.

The hearing aids were set to DSL 4.1 targets and had three separate programs:

1. Multichannel WDRC with an expansion kneepoint of 50dB SPL (high-kneepoint condition);

2. Multichannel WDRC with an expansion kneepoint of 30dB SPL (low-kneepoint condition);

3. Linear amplification with output compression limiting (control condition).

Expansion ratio was constant at 0.7:1, which represents a typical expansion ratio currently available in hearing aids.

Eight CV nonsense syllables, four voice and four unvoiced, from the Nonsense Syllable Test (Dubno & Dirks, 1982) were presented to subjects at 50, 60 and 71dB SPL. Recordings of aided stimuli were measured at the tympanic membrane for each subject (Souza & Tremblay, 2006) and signal audibility was determined using the Aided Audibility Index (AAI) (Stelmachowicz et al., 1994) using modifications for hearing-impaired subjects as described by Souza and Turner (1998).

Three of Brennan and Souza's (2009) hypotheses were confirmed: high-expansion kneepoints significantly reduced signal audibility for speech at all presentation levels and consonant-vowel-recognition scores were significantly lower for the high-kneepoint condition, especially at presentation levels of 50dB and 60dB SPL. Subsequent regression analyses revealed that CV syllable-recognition scores were significantly associated with audibility. The authors' presumption that the effect of expansion on audibility and speech recognition would decrease with increasing speech-presentation levels was not confirmed. Instead, expansion had negative effects on CV recognition and audibility at all presentation levels. This was in contrast with previous work reporting that expansion did not affect speech recognition above certain levels (Bray and Ghent, 2001; Plyler et al., 2005, 2007; Walker et al., 1984), but this discrepancy may be explained by differences in presentation level, speech materials, expansion ratio, time constants, or other hearing aid settings.

Despite some variability in results across studies, it is clear that high-expansion kneepoints result in decreased speech-recognition scores, presumably due in part to decreased audibility. Other potential explanations involve degradation of temporal cues and disruption of the intensity relationships between consonants and vowels, which provide important cues for consonant recognition (Hedrick & Younger, 2007; Walker et al., 1984).

Expansion is a feature of modern hearing aids that is often misunderstood; because it is characterized in terms of a ratio and kneepoint, it may be easily confused with compression. Essentially the opposite of compression, expansion results in less amplification for softer sounds than louder sounds. The expansion kneepoint is often the same as the compression kneepoint, indicating that expansion occurs below the kneepoint level and compression occurs above it. Alternatively, the input/output function of a circuit might show a region of linearity between the expansion and compression kneepoints.

Regardless of its various characteristics, expansion may help reduce the perception of unwanted environmental and hearing aid circuit noise, resulting in improved subjective hearing aid performance. However, because audibility and speech recognition are two primary goals of amplification, it is essential to ensure that expansion does not result in decreased objective performance. Brennan and Souza (2009) suggest that the use of active noise reduction for lower-level stimuli may provide the benefits of expansion without the negative effect on speech recognition, but more research on this topic is warranted. Because expansion is commonly used in current hearing instruments, it is important for audiologists to understand the principles of compression and expansion so that appropriate settings can be selected to maximize audibility and comfort for individual hearing aid users. And as always, counseling is essential for preparing both new and experienced hearing aid users for adjustment to new hearing aid technology and the perception of normal environmental sounds.

References

Brennan, M., & Souza, P. (2009). Effects of expansion on consonant recognition and consonant audibility. *Journal of the American Academy of Audiology*, 20, 119–127.

Bray, V.H. & Ghent, R.M. (2001). Expansion as a tool for adaptation to amplification. *Seminars in Hearing*, 22, 183–198.

Dubno, J.R., & Dirks, D.D. (1982). Evaluation of hearing-impaired listeners using a Nonsense-Syllable Test: I test reliability. *Journal of Speech, Language and Hearing Research*, 25, 135–141.

Hedrick, M.S., & Younger, M.S. (2007). Perceptual weighting of stop consonant cues by normal and impaired listeners in reverberation versus noise. *Journal of Speech, Language and Hearing Research*, 50, 254–269.

Plyler, P., Hill, A., & Trine, T. D. (2005). The effects of expansion on the objective and subjective performance of hearing instrument users. *Journal of the American Academy of Audiology, 16*, 101–113.

Plyler, P.N., Lowery, K.J., Hamby, H.M., & Trine, T.D. (2007). The objective and subjective evaluation of multichannel expansion in wide dynamic range compression hearing instruments. *Journal of Speech, Language and Hearing Research*, 50, 15–24.

Souza, P.E, & Tremblay, K.L. (2006). New perspectives on assessing amplification effects. *Trends in Amplification*, 10, 119–143.

Souza, P.E., & Turner, C.W. (1998). Multichannel compression, temporal cues and audibility. *Journal of Speech, Language and Hearing Research*, 41, 315–326.

Stelmachowicz, P.G., Lewis, D.E., Kalberer, L., & Creutz, T. (1994). *Situational Hearing Aid Response Profile User's Manual (SHARP, v 6.0)*. Omaha: Boys Town National Research Hospital.

Walker, G., Byrne, D., & Dillon, H. (1984). The effects of multichannel compression/expansion amplification on the intelligibility of nonsense syllables in noise. *Journal of the Acoustical Society of America*, 76, 746–757.

Zakis, J.A., & Wise, C. (2007). The acoustic and perceptual effects of two noise suppression algorithms. *Journal of the Acoustical Society of America*, 121, 433–441.

Some Benefits of Increasing the Number of Compression Channels

Yund, W.E., & Buckles, K.M. (1995). Multichannel compression hearing aids: Effect of number of channels on speech discrimination in noise. *Journal of the Acoustical Society of America,* 97(2), 1206–1223.

The Yund and Buckles (1995) study of multichannel compression hearing aids offers valuable insight for the recommendation and fitting of current hearing aids. A wide array of hearing aids is available, at different technology levels, and the number of compression channels is often a distinguishing feature. Typically, audiologists may recommend higher quality or more sophisticated processing for patients who lead active lives and participate in activities in challenging listening environments. The effect of multiple compression channels on hearing aid performance in noise is therefore an important consideration in determining the most appropriate instrument for an individual patient.

There are theoretical reasons to expect multiple channels of compression to improve or reduce speech-recognition ability in the presence of noise. Potential benefits of multichannel compression include plausible reduction of competing noise and preservation of audible speech information, as well as better fit to audiometric thresholds. Because the amplification within a given channel is based on the signal-to-noise ratio within that channel, it follows that speech information could be lost due to inadequate application of gain in instruments with fewer channels. Most previous studies of multichannel compression hearing aids compared their performance to that of linear hearing aids, but a study by Barfod (1978) showed improvement in performance as the number of channels increased from two to four. At the time (1995) the reviewed paper was published, there were few studies to support performance increases with an increasing number of compression channels.

Yund and Buckles (1995) acknowledged potential detrimental effects of multichannel compression, including reduction of temporal and spectral contrasts, resulting in increased phoneme confusion (Bustamante & Braida, 1987; Plomp, 1988, 1989; Villchur, 1989). The authors point out that, theoretically, any deleterious effects of multichannel compression should increase with increasing number of channels because systems with more channels would be better able to detect and respond to intensity variations across frequency. They suggest that at some point any benefit in multiple channel compression should plateau and possibly begin to decrease with increasing channels. Some have argued, however, that a reduction in amplitude contrasts might be counterbalanced somewhat by recruitment effects, which could result in larger perceptual contrasts (Moore, 1991).

Sixteen subjects with sensorineural hearing loss of varying degrees and etiologies participated in the experiment. The authors included participants with a broad range of hearing losses in an effort to examine any interactions between compression characteristics and hearing-loss configuration or severity. Test materials were based on the Nonsense Syllable Test (NST) by Resnick et al. (1975). The original test used nonsense syllables presented in a carrier phrase, but the present study used newly recorded stimuli presented in isolation. Test stimuli were CV or VC monosyllables that varied according to voicing (voiced or voiceless), consonant position (initial or final), and vowel context (/a/, /i/, or /u/). All stimuli were recorded by a male speaker and a female speaker.

The nonsense syllables were combined with speech-shaped noise (French & Steinberg, 1947) prior to any linear or multichannel compression processing. The intensity of the noise was always 70dB SPL, and speech was added to noise at 85, 80, 75, 70, and 65dB SPL to yield signal-to-noise ratios ranging from +15dB to -5dB. Stimuli for each talker and each signal-to-noise ratio were presented in seven different processing conditions: unprocessed, linear amplification (20 dB, flat), four-channel compression, six-channel compression, eight-channel compression, twelve-channel compression, and sixteen-channel compression.

Not surprisingly, the results showed a strong effect of signal-to-noise ratio, with performance deteriorating as intensity of the speech signal decreased. In quiet conditions, discrimination was better for the male voice than for the female voice, but this reversed in the presence of noise, resulting in an advantage for discrimination of the female voice. The average spectrum for the female voice had relatively more energy in the high frequencies than the male voice and had peaks occurring at higher frequencies. The authors pointed out that because speech-spectrum noise has less energy in high frequencies, it may be less effective at masking the female voice, thus causing a reduced performance decrement for the female voice in the presence of noise.

There was a significant interaction between number of channels and voice; increasing the number of channels produced a greater improvement in discrimination for the male than it did for the female voice. As noted above, speech-weighted noise has more energy in low frequencies, which could be more effective at masking the male voice. Because gain reductions in a multi-channel hearing aid are based on the level within a given channel, an instrument with fewer channels would be expected to lose more low-frequency speech information in its attempts to reduce gains applied to high-level, low-frequency noise. Instruments with a higher number of processing channels would therefore be expected to have less of a detrimental effect on low-frequency speech information, which is exactly what the results of the current study suggest.

There was a significant effect overall for the number of processing channels. Speech discrimination clearly improved as the number of channels increased from four to eight channels and remained consistent above eight channels. There was no significant interaction between number of channels and signal-to-noise ratio; increases in the number of channels did not yield further improvement or decrement for performance at poor signal-to-noise ratios.

The authors conducted a detailed analysis of consonant discrimination, which yielded information about the transmission of specific stimulus features and how they changed with increasing channels. In general, voiceless consonants were discriminated better than voiced, CV monosyllables were discriminated better than VC, and consonants in the context of /a/ were discriminated better than those in the context of /i/ or /u/. None of these features varied significantly as the number of channels increased. There was, however, a differential effect of number of channels on the perception of place and manner of articulation. Increasing the number of channels yielded more improvement for middle consonants than front or back consonants, and improved fricative perception than nasals and glides, and voiceless stops more than voiced stops. The most striking characteristic to benefit from increased channels was duration, which is a particularly important cue for differentiating fricatives.

The authors analyzed the frequency responses of the multi-channel instruments in their study and found that overall they were remarkably similar with one notable exception. Average amplification at 4 kHz and above increased as the number of processing channels increased from four to eight to sixteen channels. The improved high-frequency response for instruments with more channels of processing, resulting in better transmission of high-frequency speech cues, is likely to help account for the noted improvements in consonant discrimination.

One aim of the current study was to determine any negative effects of increasing channels of compression. They found no negative effects, at least up to sixteen channels, which was the maximum number used in the study. The authors mentioned a few contemporary studies that found increasingly negative effects with increasing number of channels for normal and hearing-impaired subjects, but only when high-compression ratios were used (greater than 3:1). It appears that for the compression characteristics used in the current study, any potential negative effect of increasing compression channels was negated by increases in the availability of speech information or possibly, as Moore (1995) suggested, recruitment effects.

Audiologists usually make recommendations for hearing aid style based on audiometric configuration, manual dexterity, and anatomical variables, but the choice of technology level is often based on a patient's lifestyle. There may be many reasons to recommend premium instruments for patients with active lifestyles, including more effective directional microphones, better automatic processing, and more precise programming adjustments. The current study supports the importance of multiple channels of processing for better performance in noise, which is one of the major considerations for hearing aid users who participate in activities in challenging listening environments. Specifically, their study showed benefits for instruments having up to eight channels of processing, a delineation that differentiates many entry-level hearing aids from their more sophisticated counterparts.

Yund and Buckles (1995) used multichannel compression instruments that were simpler than those available today. Advances in digital processing have led to instruments that vary with regard to speed of processing, compression characteristics, adaptive directionality, noise reduction, and other parameters. Although an updated investigation of their hypotheses with current hearing aid technology could provide interesting insights into their findings, their study still supports the potential benefit of increased channels of processing for individuals with a wide range of hearing losses.

References

Barfod, J. (1978). Multichannel compression hearing aids: Experiments and considerations on clinical applicability. In C. Ludvigsen and J. Barfod (Eds.). *Sensorineural Hearing Impairment and Hearing Aids, Scandinavian Audiology, Supplementum,* 6, 315–340.

Bustamante, D.K., & Braida, L.D. (1987). Principal-component amplitude compression for the hearing impaired. *Journal of the Acoustical Society of America,* 82, 1227–1242.

French, N.R., & Steinberg, J.C. (1947). Factors governing the intelligibility of speech sounds. *Journal of the Acoustical Society of America,* 19, 90–119.

Moore, B.C.J. (1991). Characterization and simulation of impaired hearing: Implications for hearing aid design. *Ear and Hearing,* 12, 154S–161S.

Plomp, R. (1988). The negative effect of amplitude compression in multichannel hearing aids in the light of the modulation-transfer function. *Journal of the Acoustical Society of America,* 83, 2322–2327.

Plomp, R. (1989). "Reply to 'Comments on 'The negative effect of amplitude compression in multichannel hearing aids in the light of the modulation-transfer function. [*Journal of the Acoustical Society of America* 83, 2322-2327.]' [*Journal of the Acoustical Society of America,* 86, 425-427]." *Journal of the Acoustical Society of America,* 86, 428.

Resnick, S.B., Dubno, J.R., Hoffnung, S. & Levitt, H. (1975). Phoneme errors on a nonsense syllable test. *Journal of the Acoustical Society of America,* 58, 114(A).

Villchur, E. (1989). "Comments on 'The negative effect of amplitude compression in multichannel hearing aids in the light of the modulation-transfer function. ' [*Journal of the Acoustical Society of America* 83, 2322-2327.]." *Journal of the Acoustical Society of America,* 86, 425–427.

Yund, W.E., & Buckles, K.M. (1995). Multichannel compression hearing aids: Effect of number of channels on speech discrimination in noise. *Journal of the Acoustical Society of America,* 97(2), 1206–1223.

Are You Prescribing an Appropriate MPO?

Kuk, F., Peeters, H., Korhonen, P., & Lau, C. (2010). Effect of MPO and noise reduction on speech recognition in noise. *Journal of the American Academy of Audiology*, submitted November 2010.

A component of clinical best practice would suggest that audiologists determine a patient's uncomfortable listening levels in order to prescribe the output limiting characteristics of a hearing aid (Hawkins et al., 1987). The optimal maximum power output (MPO) should be based on two goals: preventing loudness discomfort and avoiding distorted sound quality at high levels. The upper limit of a prescribed MPO must allow comfortable listening; less consideration is given to the consequences that under prescribing MPO might have on hearing aid and patient performance.

There are two primary concerns related to setting the acceptable lower MPO limit: saturation and insufficient loudness. Saturation occurs when the input level of a stimulus plus gains applied by the hearing aid exceed the MPO, causing distortion and temporal smearing (Dillon & Storey, 1998). This results in a degradation of speech cues and a perceived lack of clarity, particularly in the presence of competing noise. Similarly, insufficient loudness reduces the availability of speech cues. There are numerous reports of subjective degradation of sound when MPO is set lower than prescribed levels, particularly in linear hearing instruments (Kuk et al., 2008; Preminger, et al., 2001; Storey et al., 1998). There is not yet consensus on whether low MPO levels also cause objective degradation in performance.

The purpose of the study described here was to determine if sub-optimal MPO could affect speech intelligibility in the presence of noise, even in a multi-channel, nonlinear hearing aid. Furthermore, the authors examined if gain reductions from a noise-reduction algorithm could mitigate the detrimental effects of the lower MPO. The authors reasoned that a reduction in output at higher input levels, via compression and noise reduction, could reduce distortions that arise from output limiting.

Eleven adults with flat, severe hearing losses participated in the reviewed study. Subjects were fitted bilaterally with fifteen-channel, wide dynamic range compression, behind-the-ear hearing aids. Microphones were set to omnidirectional and other than noise reduction, no special features were activated during the study. Subjects responded to stimuli from the Hearing in Noise Test (HINT) (Nilsson et al., 1994) presented at a 0-degree azimuth angle in the presence of continuous speech-shaped noise. The HINT stimuli yielded reception thresholds for speech (RTS) scores for each test condition.

Test conditions included two MPO prescriptions: the default MPO level (Pascoe, 1989) and 10dB below that level. The lower setting was chosen based on previous work that reported an approximately 18dB acceptable MPO range for listeners with severe hearing loss (Storey et al., 1998). MPOs set at 10dB below default would therefore be likely to approach the low end of the acceptable range, resulting in perceptual consequences. Speech-shaped noise was presented at two levels: 68dB SPL and 75dB SPL. Testing was done with and without digital noise reduction (DNR).

Analysis of the HINT RTS scores yielded significant main effects of MPO and DNR, as well as significant interactions between MPO and DNR, and DNR and noise level. There was no significant difference between the two noise-level conditions. Subjects performed better with the default MPO setting versus the reduced MPO setting. The interaction between the MPO and DNR showed that subjects' performance in the low-MPO condition was less degraded when DNR was activated. These findings support the authors' hypotheses that reduced MPO can adversely affect speech discrimination and that noise-reduction processing can at least partially mitigate these adverse effects.

Prescriptive formulae have proven to be reasonably good predictors of acceptable MPO levels (Preminger et al., 2001; Storey et al., 1988). In contrast, there is some question as to the value of clinical UCL testing prior to fitting, especially when validation with loudness measures is performed after the fitting (Mackersie, 2006). Improper instruction for the UCL task may yield inappropriately low

UCL estimates, resulting in compromised performance and sound quality. The authors of the current paper recommend following prescriptive targets for MPO and conducting verification measures after the fitting, such as real-ear saturation response (RESR) and subjective loudness judgments.

Another scenario, and an ultimately avoidable one, involves individuals who have been fitted with inappropriate instruments for their loss, usually because of cosmetic concerns. It is unfortunately not so unusual for some individuals with severe hearing losses to be fitted with RIC or CIC instruments because of their desirable cosmetic characteristics. Smaller receivers will likely have MPOs that are too low for hearing aid users with severe hearing loss. Many patients may not realize they are giving anything up when they select a CIC or RIC and may view these styles as equally appropriate options for their loss. The hearing aid selection process must therefore be guided by the audiologist; patients should be educated about the benefits and limitations of various hearing aid options and counseled about the adverse effects of under-fitting their loss with a more cosmetically appealing option.

The results of the current study are important because they illuminate an issue related to hearing aid output that might not always be taken into clinical consideration. MPO settings are usually thought of as a way to prevent loudness discomfort, so the concern is to avoid setting the MPO too high. Kuk and colleagues (2010) have shown that an MPO that is too low could also have adverse effects and have provided valuable information to help clinicians select appropriate MPO settings. Additionally, their findings show objective benefits and support the use of noise-reduction strategies, particularly for individuals with reduced dynamic range due to severe hearing loss or tolerance issues. Of course their findings may not be generalizable to all multi-channel compression instruments, with the wide variety of compression characteristics that are available, but they present important considerations that should be examined in further detail with other instruments.

References

Dillon, H., & Storey, L. (1998). The National Acoustic Laboratories' procedure for selecting the saturation sound pressure level of hearing aids: theoretical derivation. *Ear and Hearing*, 19(4), 255–266.

Hawkins, D., Walden, B., Montgomery, A., & Prosek, R. (1987). Description and validation of an LDL procedure designed to select SSPL90. *Ear and Hearing*, 8, 162–169.

Kuk, F., Korhonen, P., Baekgaard, L., & Jessen, A. (2008). MPO: A forgotten parameter in hearing aid fitting. *Hearing Review* 15(6), 34–40.

Kuk, F., Peeters, H., Korhonen, P., & Lau, C. (2010). Effect of MPO and noise reduction on speech recognition in noise. *Journal of the American Academy of Audiology*, submitted November 2010.

Mackersie, C. (2006). Hearing aid maximum output and loudness discomfort: are unaided loudness measures needed? *Journal of the American Academy of Audiology*, 18 (6), 504–514.

Nilsson, M., Soli, S., & Sullivan, J. (1994). Development of the Hearing in Noise Test for the measurement of speech reception thresholds in quiet and in noise. *Journal of the Acoustical Society of America*, 95(2), 1085–1099.

Pascoe, D. (1989). Clinical measurements of the auditory dynamic range and their relation to formulae for hearing aid gain. In J. Jensen (Ed.), *Hearing Aid Fitting: Theoretical and Practical Views. Proceedings of the 13th Danavox Symposium*, 129–152. Copenhagen: Danavox.

Preminger, J., Neuman, A., & Cunningham, D. (2001). The selection and validation of output sound pressure level in multichannel hearing aids. *Ear and Hearing*, 22(6), 487–500.

Storey, L., Dillon, H., Yeend, I., & Wigney, D. (1998). The National Acoustic Laboratories' procedure for selecting the saturation sound pressure level of hearing aids: experimental validation. *Ear and Hearing*, 19(4), 267–279.

The Patient with Hearing Loss

What Motivates Hearing Aid Use?

Jenstad, L. & Moon, J. (2011). Systematic review of barriers and facilitators to hearing aid uptake in older adults. *Audiology Research*, 1:e25, 91–96.

Though some causes of adult-onset hearing loss are treated medically or surgically, hearing aid use is by far the most common treatment. Yet 25% of adults who could benefit from hearing instruments actually wear them (Kochkin, 2000; Meister et al., 2008). A number of studies have examined the factors that prevent individuals from purchasing hearing aids and Jenstad and Moon's (2011) objective was to systematically review the literature to identify the main barriers to hearing aid uptake in older adults.

They included subjective and objective reports, but limited this investigation to studies with more than fifty subjects over the age of sixty-five, who had never used hearing aids, had at least mild-to-moderate sensorineural hearing loss, and were in good general health. From an initial set of 388 abstracts, they eliminated studies about children, cochlear implants, medical aspects of hearing loss, auditory processing, or hearing aid outcomes. From the remaining fifty articles, the report focused on fourteen papers that met the inclusion criteria. Hearing aid uptake was defined as a hearing aid purchase, but some studies included willingness to purchase. Based on the literature review, Jenstad and Moon (2011) identified a set of predictors of hearing aid uptake in older adults. Some of the predictors they described may be helpful discussion points for audiologists counseling potential hearing aid users.

Self-reported hearing loss was evaluated in questionnaires and hearing-handicap indices that examined hearing-related quality of life as well as activity and participation limitation (Chang et al., 2009; Garstecki & Erler, 1998; Helvik et al., 2008; Meister et al., 2008; Palmer, 2009). Not surprisingly, as self-reported hearing loss increased, study participants were more willing to obtain hearing aids.

In other words, the more aware individuals were of their hearing-related difficulties, the more likely they were to purchase hearing aids. With this in mind, audiologists should instruct unmotivated hearing aid candidates to pay close attention to their hearing-related difficulties while determining need for amplification. Work together to identify activities that patients must do (e.g., work) or enjoy doing (e.g., dining out, going to the—movies or theater). Using this information to understand the extent to which hearing loss disrupts their communication ability in these situations will enlighten the patient to the extent of their own hearing handicap and point towards opportunities for treatment and counseling.

Stigma was predictive of hearing aid acceptance in some studies (Franks & Beckmann, 1985; Garstecki & Erler, 1998; Kochkin, 2007; Meister et al., 2008; Wallhagen, 2010), but overall was inconsistent in its effect on hearing aid uptake. In 1985, Franks and Beckmann found that stigma was the highest concern among their subjects, whereas in 2008, Meister and associates found that stigma only accounted for 8% of the variability in hearing aid uptake. The negative stigma associated with hearing aids is assumed to relate to the appearance of the aid and the perception of hearing loss by other people. Therefore, hearing aid users with high concern desire small, discreet instruments. Improvements in technology allow for smaller, sleeker designs that make the hearing aid—and hearing loss—less noticeable. Therefore, hearing aid users no longer have to make an obvious acknowledgement of their hearing impairment.

Degree of hearing loss was a significant factor in the decision to obtain hearing aids, but the effect seems to be modified by gender. Garstecki and Erler (1998) found that degree of hearing loss was more likely to affect hearing aid uptake for females than males, but this finding was not reported in other studies. In general, as degree of hearing loss worsens, people are more willing to wear hearing aids. Detailed discussion of audiometric findings, with visual references to speech and environmental sound levels, helps to familiarize the patient and their family with degrees of hearing loss and the impact on speech perception. Using tools like hearing-loss simulators offer a convenient tool for educating and motivating patients toward the acceptance of hearing aids.

Personality and psychological factors affected hearing aid uptake in three studies (Cox et al., 2005; Garstecki & Erler, 1998; Helvik et al., 2008). Cox and her colleagues found that hearing aid "seekers" were less neurotic, more open, and more agreeable than those who did not seek hearing aids. Internal locus of control predicted hearing aid acceptance in Cox's study (2005), but Garstecki and Erler (1988) found that it was only predictive for female subjects. Though locus of control is one among many factors influencing the decision, the choice to obtain hearing aids should be presented as a way to assume control of the hearing impairment and make proactive steps toward improving communication abilities.

Helvik et al. (2008) found that subjects who reported using fewer maladaptive coping strategies, such as dominating conversations or avoiding social situations, were less likely to accept hearing aids. Many hearing-impaired individuals use poor coping strategies without realizing it. It seems counterintuitive that reported use of maladaptive strategies would be inversely related to hearing aid acceptance, but the authors surmised that the study participants who rejected hearing aids may have been in denial about both the hearing loss and their use of poor communication strategies. Hearing impaired individuals may not be aware of the extent of their communication difficulties and may not realize how often they are misunderstanding conversation or requiring others to make extra efforts. Including family members in the discussion of hearing aid candidacy is critical to make the hearing-impaired individual aware of how their loss affects others and how the use of poor or ineffective strategies may result in frustration for themselves and other conversational participants.

Cost was a barrier to hearing aid use in some studies but was not a significant factor in others (Meister et al., 2008). But Jenstad and Moon point out that cost may affect hearing aid acceptance in more than one way. Kochkin's 2007 survey found that 64% of respondents reported that they could not afford hearing aids, whereas 45% of respondents said that hearing aids are not worth the expense. There are ways in which audiologists can address both of these issues with hearing aid candidates. First, improvements in technology have made quality instruments available at a wide range of prices. Most manufacturers offer a broad product line, with entry-level instruments in custom and BTE styles. Patients should be assured that their hearing loss, lifestyle, and listening needs will determine a range of options from which to choose. Lower-cost hearing aids might require more manual adjustment than aids with sophisticated automatic features, but with proper training and programming, some lower-cost options might work quite well. Additionally, unbundled pricing and financing options may help potential hearing aid users better understand the total purchase price. Together, these strategies make cost less of a barrier for many potential hearing aid candidates.

Kochkin's (2007) finding that 45% of respondents felt hearing aids were not worth the expense is perhaps more difficult to address. Some of the bias against hearing aids is related to inappropriate hearing aid selection or inadequate training and follow-up care. Most audiologists have encountered patients with a friend or neighbor who doesn't like their hearing aids. Negative experiences with hearing aids may be more likely related to selection, programming, and follow-up care than the quality of the hearing instruments themselves. The finest hearing aid available will be rejected if it is inappropriate for the user's hearing loss or lifestyle or is programmed improperly. Unfortunately, many people who have an unsuccessful experience have acquired their hearing aids through non-clinical channels. These people often blame their dissatisfaction on the quality of the hearing aid, contributing to a larger general perception that hearing aids are not worth the price. Patients must emphasize the importance of the care that they provide. Thorough verification, validation, and follow-up care by well-trained audiologists will affect patient's perception and lead them toward success.

The effect of age on hearing aid uptake was unclear in Jenstad and Moon's review. One study showed a slight increase in hearing aid uptake with increasing age (Helvik et al., 2008), whereas another showed a stronger increase with age (Hidalgo et al., 2009). In contrast, Uchida et al. (2008) found that hearing aid uptake decreased with increasing age. The effect of age, if any, on hearing aid acceptance will be confounded by other variables such as degree of loss, lifestyle, general health, and financial constraints. Therefore, age should be a minor consideration with reference to hearing aid candidacy but remains highly relevant when discussing specific options such as manual controls, automatic features, and hearing aid styles.

Gender affected the predictive value of several factors including stigma, degree of loss, and locus of control. Hidalgo et al. (2009) found that in general males were more likely to report a need for hearing aids than were females. Gender in itself might not be a strong predictor, so it probably should not be specifically considered in discussions with potential hearing aid users, as other variables appear to have more impact on the decision to pursue hearing aids.

Franks and Beckmann (1985) reported that individuals who chose not to purchase hearing aids were more likely to report that hearing aids were inconvenient to wear. Though the study was done in 1985, their findings merit consideration today. Since then, hearing aids have become smaller, more effective, and less troublesome because of advances such as feedback cancellation, directional microphones, and noise reduction. However, the fact remains that hearing aids must be worn, cleaned, and cared for daily, and in most cases batteries must be changed on a weekly basis. Use and care guidelines should be balanced by discussion of the likely benefits of hearing aid use and the positive effect they have on communication in everyday situations. With the technological sophistication that today's hearing aids offer, the known benefits should outweigh any perceived inconvenience.

Jenstad and Moon (2011) have clarified some of the primary barriers to hearing aid uptake, providing useful information for clinicians working with hearing aid candidates. The predictors they discussed can be addressed systematically to quell concerns about and underscore the need for hearing instruments. Discussing these issues at the outset may encourage motivated clients to proceed with a hearing aid purchase and provide helpful considerations for those who are not yet ready to pursue amplification. With many potential places to purchase and limited information to guide patients toward qualified hearing care professionals, Internet sales offer the appealing promise of quality hearing instruments at lower costs than may be found in a clinic. But consumers must be educated that a key to successful hearing aid use is the support of the professional, not the quality of the device itself. Anyone can "sell" a quality hearing aid, but only a trained professional can make appropriate clinical decisions and recommendations.

References

Chang, H.P., Ho, C.Y., & Chou, P. (2009). The factors associated with a self-perceived hearing handicap in elderly people with hearing impairment—results from a community-based study. *Ear and Hearing*, 30(5), 576–583.

Cox, R.M., Alexander, G.C., & Gray, G.A. (2005). Who wants a hearing aid? Personality profiles of hearing aid seekers. *Ear and Hearing*, 26(1), 12–26.

Franks, J.R. & Beckmann, N.J. (1985). Rejection of hearing aids: attitudes of a geriatric sample. *Ear and Hearing,* 6(3), 161–166.

Garstecki, D.C., & Erler, S. F. (1998). Hearing loss, control and demographic factors influencing hearing aid use among older adults. *Journal of Speech, Language and Hearing Research,* 41(3), 527–537.

Helvik, A.S., Wennberg, S., Jacobsen, G., & Hallberg, L.R. (2008). Why do some individuals with objectively verified hearing loss reject hearing aids? *Audiological Medicine,* 6(2), 141–148.

Hidalgo, J.L., Gras, C.B., Lapeira, J.T., Verdejo, M.A., del Campo, D.C., & Rabadan, F.E. (2009). Functional status of elderly people with hearing loss. *Archives of Gerontology and Geriatrics,* 49(1), 88–92.

Jenstad, L., & Moon, J. (2011). Systematic review of barriers and facilitators to hearing aid uptake in older adults. *Audiology Research,* 1:e25, 91–96.

Kochkin, S. (2000). MarkeTrak V: "Why my hearing aids are in the drawer": the consumer's perspective. *Hearing Journal,* 53(2), 34–41.

Kochkin, S. (2007). MarkeTrak VII: Obstacles to adult non-user adoption of hearing aids. *The Hearing Journal,* 60(4), 24–50.

Meister, H., Walger, M., Brehmer, D., von Wedel, U., & von Wedel, J. (2008). The relationship between pre-fitting expectations and willingness to use hearing aids. *International Journal of Audiology,* 47(4), 153–159.

Uchiga, Y., Sugiura, S., Ando, F., Shimokata, H., Yoshioka, M. & Nakashima, T. (2008). Analyses of factors contributing to hearing aid use and both subjective and objective estimates of hearing. *Nippon Jibiinkoka Gakkai Kaiho,* 111(5), 405–411.

Wallhagen, M.I. (2010). The stigma of hearing loss. *Gerontologist,* 50(1), 66–75.

Palmer, C.V., Solodar, H.S., Hurley, W.R., Byrne, D.C. & Williams, K.O. (2009). Self-perception of hearing ability as a strong predictor of hearing aid purchase. *Journal of the American Academy of Audiology,* 20(6), 341–347.

Does Listening Experience Affect a Hearing Aid Wearer's Preferred Gain?

Keidser, G., O'Brien, A., Carter, L., McLelland, M., & Yeend, I. (2008). Variation in preferred gain with experience for hearing aid users. *International Journal of Audiology*, 47, 621–635.

Auditory plasticity, or the reorganization of neural connections in the auditory system, has been documented in studies with human and animal subjects (Palmer et al., 1998; Philibert et al., 2005; Willott, 1996). These studies found that the representation of acoustic stimuli along auditory pathways can change based on auditory experience. The concept of auditory plasticity may relate to hearing aid patients in two ways: first, neural reorganization is likely to occur in response to the hearing loss itself, then subsequent reorganization may occur in response to amplification. Indeed, most audiologists observe that new hearing aid users experience an initial adjustment period in which they prefer less gain, but that over time they are able to accept increases in gain.

The clinical manifestation of auditory plasticity is sometimes associated with acclimatization to amplification and has been studied in hearing aid users in numerous ways: preferred gain for new and experienced users (Cox and Alexander, 1992; Horwitz & Turner, 1997; Marriage et al., 2004), speech performance over time (Bentler et al., 1993a; Gatehouse, 1992), subjective benefit or sound quality over time (Bentler et al., 1993b; Ovegard et al., 1997), loudness perception and intensity discrimination over time (Olsen et al., 1999; Philibert et al., 2002), even changes in ABR wave V latency (Philibert et al., 2005). Most studies have found small but significant changes over time as participants adjusted to amplified sound. Others, however, have found no significant difference between new and experienced hearing aid users (Smeds et al., 2006a, 2006b). Munro and Lutman (2003) suggest that acclimatization may occur specifically in relation to high-level, high-frequency sounds.

The authors of the current study were specifically interested in comparing gain preferences and loudness perception in new hearing aid users and users with more than three years of experience with hearing aids. Fifty new users and twenty-six experienced users, most with mild-to-moderate sensorineural hearing loss, were fitted with digital, two-channel, WDRC instruments equipped with volume controls. Compression attack and release characteristics were set identically for all subjects and a noise-reduction algorithm was turned off. The hearing aids had three independent programs:

1.   NAL-NL1 target response
2.   NAL-NL1 with a 6dB high-frequency cut at 3000Hz (HFC)
3.   NAL-NL1 with a 6dB low-frequency cut at 500Hz (LFC)

Subjects were asked to compare the programs in everyday environments and record their preferred overall program. Follow-up testing was conducted at one month, four months and thirteen months post-fitting and subjects were specifically instructed to arrive at each appointment with the device on their overall preferred program and volume control setting. At each appointment, 2-cc coupler and real-ear measurements were obtained with hearing instruments at the preferred settings. Questionnaires were administered to record hearing aid usage time, ranking and performance of the three programs, and to what extent the volume control was used. Loudness perception tests were performed using a categorical loudness scale test (Cox & Alexander, 1997) to determine the median SPL levels that were categorized as "comfortable."

The authors found that new and experienced users preferred the high-frequency cut (HFC) program most often. Initially, about 60% of the new users preferred the HFC program, but by thirteen months post-fitting, the preferences of new and experienced users were very similar, with approximately half of the subjects still preferring the HFC program. Fewer than 10% of the users preferred the LFC program across the duration of the study.

On average, overall preferred gain was 3dB lower for new users and increases were noted at subsequent appointments. By the time of the final appointment, new users reported higher gain settings than they did before, but did not reach the preferred levels of experienced users. This suggests that the gain acclimatization process for some users may continue beyond the thirteen-month point. Degree of hearing loss had a significant effect, as subjects with moderate hearing loss preferred 6dB lower overall gain than those with mild hearing loss.

The findings of Keidser and colleagues (2008) offer important implications for clinical practice. First, it appears that new hearing aid users experience acclimatization with regard to comfortable loudness and preferred gain settings. This supports the use of adaptation levels in hearing aid software, though some adaptation managers may provide larger decrements in overall gain (5-10dB) than many hearing aid users require. The fact that new users with mild hearing loss did not prefer as much initial gain reduction as those with moderate losses indicates that audiometric thresholds should be considered. The authors noted that more information is needed about acclimatization effects in new hearing aid users with more than moderate hearing loss. However, it is probably appropriate for audiologists to assume that for patients with moderate-to-severe hearing loss, lower initial gain settings may be needed and an extended period of adjustment may be expected.

Recent emphasis on evidence-based clinical practices underscores the importance of verification measures to ensure adequate gain and frequency response in new hearing aids. However, studies of auditory acclimatization demonstrate that it is equally important to evaluate the patient's perception of the amplified sound to ensure satisfaction. Ultimately, a patient with excellent aided test results may still reject new hearing aids if they are not comfortable.

Hearing aid acclimatization can be measured several ways, many of which are not viable for busy clinical practices. But there are two simple ways in which it should be addressed during the hearing aid fitting and follow-up appointments. At the fitting appointment, patients should be counseled about appropriate expectations for the adjustment process. For instance, patients who know ahead of time that it is normal to notice, and perhaps be slightly annoyed by, newly amplified sounds are less likely to be disheartened when this occurs. They should be advised to wear their hearing aids as consistently as possible and to report any discomfort or pain so it can be addressed with programming adjustments.

Initially, high-frequency gain in particular may need to be reduced relative to target settings. However, care should be taken to determine each individual's comfort with high-frequency sounds. In the current study, individuals were free to reduce high-frequency gain at any time by selecting the HFC program. It seems appropriate to question whether this affected their ability to adjust to high-frequency gain and if they would have eventually been able to tolerate, even prefer, more high-frequency amplification had they not been able to switch at will into the HFC program. The importance of high-frequency information for speech intelligibility, especially in noise, is well established (Hornsby & Ricketts, 2003; Turner & Henry, 2002). To avoid detrimental effects on speech perception, high-frequency gain should approach targets as closely as possible, while still maintaining patient comfort.

At follow-up appointments, patients should be questioned in detail about their comfort and overall progress with the new aids. Formal questionnaires like the APHAB (Cox and Alexander, 1995) can be used to determine specific sound preferences so that appropriate adjustments can be made. The more precise information an audiologist obtains from a patient, the more likely they are to zero in on necessary programming changes. One important point to note is that this study evaluated gain preferences only. Most hearing instruments have several adjustable parameters, and some new users might respond as favorably to increased compression ratios or lowered compression kneepoints, thereby reducing louder sounds but maintaining gain for low- to moderate-level sounds.

Because it appears that the acclimatization process may continue beyond a year, follow-up care after the initial trial period should be planned accordingly. It may be appropriate to schedule check-ups at four, eight, and twelve months post fitting. This way, final target settings can be approached systematically for each individual.

References

Bentler, R.A., Niebuhr, D.P., Getta, J.P., & Anderson, C.V. (1993a). Longitudinal study of hearing aid effectiveness. I. Objective measures. *Journal of Speech and Hearing Research*, 36, 808–819.

Bentler, R.A., Niebuhr, D.P., Getta, J.P., & Anderson, C.V. (1993b). Longitudinal study of hearing aid effectiveness. II. Subjective measures. *Journal of Speech and Hearing Research*, 36, 820–831.

Cox, R.M., & Alexander, G.C. (1992). Maturation of hearing aid benefit: Objective and subjective measurements. *Ear and Hearing*, 13, 131–141.

Cox, R.M., & Alexander, G.C. (1995). The Abbreviated Profile of Hearing Aid Benefit (APHAB). *Ear and Hearing*, 16, 176--186.

Cox, R.M., Alexander, G.C., Taylor, I.M., & Gray, G.A. (1997). The contour test of loudness perception. *Ear and Hearing*, 18, 388–400.

Gatehouse, S. (1992). The time course and magnitude of perceptual acclimatization to frequency responses: Evidence from monaural fitting of hearing aids. *Journal of the Acoustical Society of America*, 92, 1258–1268.

Hornsby, B.W. & Ricketts, T.A. (2003). The effects of hearing loss on the contribution of high- and low-frequency speech information to speech understanding. *Journal of the Acoustical Society of America*, 113(3), 1706–1717.

Horwitz, A.R., & Turner, C.W. (1997). The time course of hearing aid benefit. *Ear and Hearing*, 18, 1–11.

Keidser, G., O'Brien, A., Carter, L., McLelland, M., & Yeend, I. (2008). Variation in preferred gain with experience for hearing aid users. *International Journal of Audiology*, 47, 621–635.

Marriage, J., Moore, B.C., & Alcantara, J.I. (2004). Comparison of three procedures for initial fitting of compression hearing aids. III. Inexperienced versus experienced users. *International Journal of Audiology*, 43, 198–210.

Munro, K.J., & Lutman, M.E. (2003). The effect of speech presentation level on measurement of auditory acclimatization to amplified speech. *Journal of the Acoustical Society of America*, 114, 484–495.

Olsen, S.O., Rasmussen, A.N., Nielsen, L.H., & Borgkvist, B.V. (1999). Loudness perception is influenced by long-term hearing aid use. *Audiology*, 38, 202–205.

Ovegard, A., Lundberg, G., Hagerman, B., Gabrielsson, A., Bengtsson, M. (1997). Sound quality judgment during acclimatization of hearing aids. *Scandanavian Audiology*, 26, 43–51.

Palmer, C.V., Nelson, C.T., & Lindley, G.A. (1998). The functionally and physiologically plastic adult auditory system. *Journal of the Acoustical Society of America*, 103, 1705–1721.

Philibert, B., Collet, L., Vesson, J.F. & Veuillet, E. 2002. Intensity-related performances are modified by long-term hearing aid use: A functional plasticity? *Hearing Research*, 165, 142–151.

Philibert, B., Collet, L., Vesson, J.F., & Veuillet, E. 2005. The auditory acclimatization effect in sensorineural hearing-impaired listeners: Evidence for functional plasticity. *Hearing Research*, 205, 131–142.

Smeds, K., Keidser, G., Zakis, J., Dillon, H., Leijon, A. (2006a). Preferred overall loudness. I. Sound field presentation in the laboratory. *International Journal of Audiology*, 45, 12–25.

Smeds, K., Keidser, G., Zakis, J., Dillon, H., Leijon, A. (2006b). Preferred overall loudness. II. Listening through hearing aids in field and laboratory tests. *International Journal of Audiology*, 45, 12–25.

Turner, C.W., & Henry, B.A. (2002). Benefits of amplification for speech recognition in background noise. *Journal of the Acoustical Society of America*, 112, 1675–1680.

Willott, J.F. (1996). Physiological plasticity in the auditory system and its possible relevance to hearing aid use, deprivation effects, and acclimatization. *Ear and Hearing*, 17, 66S–77S.

Addressing Patient Complaints When Fine-Tuning a Hearing Aid

Jenstad, L.M., Van Tasell, D.J., & Ewert, C. (2003). Hearing aid troubleshooting based on patient's descriptions. *Journal of the American Academy of Audiology*, 14(7).

As part of any clinically robust protocol, a hearing aid fitting will be objectively verified with real-ear measures and validated with a speech-in-noise test. Fine tuning and follow-up adjustments are an equally important part of the fitting process. This stage of the routine fitting process does not follow standardized procedures and is almost always directed by a patient's complaints or descriptions of real-world experience with the hearing aids. This can be a challenging dynamic for the audiologist. Patients may have difficulty putting their auditory experience into words and different people may describe similar sound quality issues in different ways. Additionally, there may be several ways to address any given complaint and a given programming adjustment may not have the same effect on different hearing aids.

Hearing aid manufacturers often include a fine-tuning guide or automated fitting assistant within their software to help the audiologist make appropriate adjustments for common patient complaints. There are limitations to the effectiveness of these fine-tuning guides, in that they are inherently specific to a limited range of products and the suggested adjustments are subject to the expertise and resources of that manufacturer. The manner in which a sound quality complaint is described may differ between manufacturers and the recommended adjustments in response to the complaint may differ as well.

There have been a number of efforts to develop a single hearing aid troubleshooting guide that could be used across devices and manufacturers (Gabrielsson et al., 1979; Lundberg et al., 1992; Moore et al., 1998, 1988, 1990; Ovegard et al., 1997). The first and perhaps most challenging step toward this goal has been to determine the most common descriptors that patients use for sound quality complaints. Moore (1998) and his colleagues developed a procedure in which responses on three ratings scales (e.g., "loud versus quiet," "tinny versus boomy") were used to make adjustments to gain and compression settings. However, their procedure did not allow for the bevy of descriptors that patients create, limiting potential utility for everyday clinical settings. Gabrielsson and colleagues, in a series of Swedish studies, (Gabrielsson, 1979; Gabrielsson et al., 1990; Gabrielsson et al., 1988) developed a set of reliable terms to describe sound quality. These descriptors have since been translated and used in English language research (Bentler et al., 1993).

As hearing instruments become more complicated with numerous adjustable parameters, and given the wide range of experience and expertise of individuals fitting hearing instruments today, an independent fine-tuning guide is an appealing concept. Jenstad and colleagues (2003) proposed an "expert system" for troubleshooting hearing aid complaints. The authors explained that expert systems "emulate the decision making abilities of human experts" (Tharpe et al., 1993). To develop the system, two primary questions were asked:

1)   What terms do hearing-impaired listeners use to describe their reactions to specific hearing aid fitting problems?

2)   What is the expert consensus on how these patient complaints can be addressed by hearing aid adjustment?

There were two phases to the project. To address question one, the authors surveyed audiologists for their reports on how patients describe sound quality with regard to specific fitting problems. To address question two, the most frequently reported descriptors from the audiologists' responses were submitted to a panel of experts to determine how they would address the complaints.

The authors sent surveys to 1934 American Academy of Audiology members and received 311 qualifying responses. The surveys listed eighteen open-ended questions designed to elicit descriptive terms that patients would likely use for hearing aid fitting problems. For example, the question "If the fitting has too much low-frequency gain..." yielded responses such as "hollow," "plugged," and "echo." The questions probed common problems related to gain, maximum output, compression, physical fit, distortion, and feedback. The survey responses yielded a list of the forty most frequent descriptors of hearing aid fitting problems, ranked according to the number of occurrences.

The list of descriptors was used to develop a questionnaire to probe potential solutions for each problem. Each descriptor was put in the context of, "How would you change the fitting if your patient reports that ___?" and twenty-three possible fitting solutions were listed. These questionnaires were completed by a panel of experts with a minimum of five years of clinical experience. Respondents could offer more than one solution to a problem and the solutions were weighted based on the order in which they were offered. There was strong agreement among experts, suggesting that their responses could be used reliably to provide troubleshooting solutions based on sound quality descriptions. The expert responses also agreed with the initial survey that was sent to the group of 1934 audiologists, supporting the validity of these response sets.

The expert responses resulted in a fine-tuning guide in the form of tables or simplified flow charts. The charts list individual descriptors with potential solutions listed below in the order in that they should be attempted. For example, below the descriptor "My ear feels plugged," the first solution is to "increase vent" and the second is to "decrease low-frequency gain," The idea is that the clinician would first try to increase the vent diameter, and if that didn't solve the problem, they would move on to the second option, decreasing low-frequency gain. If an attempted solution creates another sound quality problem, the table can be utilized to address that problem in the same way.

The authors correctly point out that there are limitations to this tool and that proposed solutions will not necessarily have the same results with all hearing aids. For instance, depending on the compressor characteristics, raising a kneepoint might increase *or* decrease the gain at input levels below the kneepoint. It is up to the audiologist to be familiar with a given hearing aid and its adjustable parameters to arrive at the appropriate course of action.

Beyond manipulation of the hearing aid itself, the optimal solution for a particular patient complaint might not be the first recommendation in any tuning guide. For instance, for the fitting problem labeled "Hearing aid is whistling," the fourth solution listed in the table is "check for cerumen." This solution appeared fourth in the ranking based on the frequency of responses from the experts on the panel. However, any competent clinician who encounters a patient with hearing aid feedback should check for cerumen first before considering programming modifications.

The expert system proposed by Jenstad and colleagues (2003) represents a thoroughly examined, reliable step toward development of a universal troubleshooting guide for audiologists. Their paper was published in 2003, so some items should be updated to suit modern hearing aids. For example, current feedback management strategies result in fewer and less challenging feedback problems. Solutions for feedback complaints might now include "calibrate feedback management system" versus gain or vent adjustments. Similarly, most hearing aids now have solutions for listening in noise that extend beyond the simple inclusion of directional microphones, so "directional microphone" might not be an appropriately descriptive solution to address complaints about hearing in noise, as the patient is probably already using a directional microphone.

Overall, the expert system proposed by Jenstad and colleagues (2003) is a helpful clinical tool; especially if positioned as a guide to help patients find the appropriate terms to describe their perceptions. However, as the authors point out, it is not meant to replace prescriptive methods, measures of verification and validation, or the expertise of the audiologist. It is the audiologist's responsibility to be informed about current technology and its implications for real-world hearing aid performance and to communicate with their patients in enough detail to understand their patients' comments and address them appropriately.

References

Bentler, R.A., Nieburh, D.P., Getta, J.P., & Anderson, C.V. (1993). Longitudinal study of hearing aid effectiveness II: subjective measures. *Journal of Speech and Hearing Research*, 36, 820–831.

Jenstad, L.M., Van Tasell, D.J., & Ewert, C. (2003). Hearing aid troubleshooting based on patient's descriptions. *Journal of the American Academy of Audiology*, 14 (7).

Moore, B.C.J., Alcantara, J.I., & Glasberg, B.R. (1998). Development and evaluation of a procedure for fitting multi-channel compression hearing aids. *British Journal of Audiology*, 32, 177–195.

Gabrielsson A. (1979). Dimension analyses of perceived sound quality of sound-reproducing systems. *Scandinavian Journal of Psychology*, 20, 159–169.

Gabrielsson, A., Hagerman, B., Bech-Kristensen, T., & Lundberg, G. (1990). Perceived sound quality of reproductions with different frequency responses and sound levels. *Journal of the Acoustical Society of America*, 88, 1359–1366.

Gabrielsson, A. Schenkman, B.N., & Hagerman, B. (1988). The effects of different frequency responses on sound quality judgments and speech intelligibility. *Journal of Speech and Hearing Research*, 31, 166–177.

Lundberg, G., Ovegard, A., Hagerman, B., Gabrielsson, A., & Brandstom, U. (1992). Perceived sound quality in a hearing aid with vented and closed earmold equalized in frequency response. *Scandinavian Audiology*, 21, 87–92.

Ovegard, A., Lundberg, G., Hagerman, B., Gabrielsson, A., Bengtsson, M., & Brandstrom, U. (1997). Sound quality judgments during acclimatization of hearing aids. *Scandinavian Audiology*, 26, 43–51.

Tharpe, A.M., Biswas, G., & Hall, J.W. (1993). Development of an expert system for pediatric auditory brainstem response interpretation. *Journal of the American Academy of Audiology*, 4, 163–171.

Is a Patient's Acceptable Noise Level (ANL) a Valid Predictor of Successful Hearing Aid Use?

Nabelek, A.K., Freyaldenhoven, M.C., Tampas, J.W., Burchfield, S.B., & Muenchen, R.A. (2006). Acceptable noise level as a predictor of hearing aid use. *Journal of the American Academy of Audiology*, 17, 626–639.

A common complaint of hearing aid users is difficulty understanding speech in the presence of background noise. This has led to recent interest in examining acceptable noise levels in hearing aid users. Previous research has not shown a correlation between measures of speech perception in noise and successful hearing aid use (Bentler et al., 1993; Humes et al., 1996), nor have outcome measures generally been useful to predict success with hearing aids (Schum, 1999). The authors of the current study sought to determine if acceptable noise level measurements can be used to predict success with hearing aids.

Acceptable noise level (ANL) is defined as the difference between the most comfortable listening level for running speech and the maximum background-noise level that a listener is willing to accept (Nabelek et al., 1991). Therefore, a lower ANL indicates a better tolerance of background noise. Nabelek and associates (2006) compared full-time, part-time, and non-users of hearing aids to examine the relationship between ANL scores, results from the Speech In Noise (SPIN) test scores (Bilger et al., 1984), and hearing aid use patterns. One of their primary goals was to investigate whether ANL scores could be used clinically to predict future success with hearing aids.

The authors recruited 191 subjects with sensorineural hearing loss whose audiometric thresholds, on average, had a sloping mild-to-severe configuration. All subjects were binaural hearing aid users, having been fitted between three months and three years prior to testing. Subjects were fitted by a variety of audiologists who were independent of the study, so the hearing aid models, features, and signal-processing strategies differed among the subjects.

ANL scores and SPIN scores were obtained with speech and noise presented through the same speaker at a 0-degrees azimuth. Following sound-field testing, subjects responded to a brief questionnaire, which assigned them to one of three groups of hearing aid users: full-time, part-time, and non-users. Full-time users were defined as those who wore their hearing aids whenever they needed them, part-time users wore their hearing aids occasionally, and non-users had completely stopped wearing their hearing aids.

The first fifty-eight subjects were tested in three sessions over the first three months of acclimatization with hearing aids. These subjects showed consistent ANL and SPIN scores over the three sessions, so the authors concluded that acclimatization to new hearing aids did not affect test results and remaining subjects were tested in one session only.

The authors found that SPIN and ANL scores were generally not affected by age, gender, or pure-tone average. SPIN scores were significantly better in the aided condition for all listener groups, regardless of hearing aid use pattern. ANL scores were not significantly different between aided and unaided conditions, but unaided and aided ANLs were strongly correlated to hours of hearing aid use and overall hearing aid use pattern. Full-time hearing aid users demonstrated significantly lower ANL scores than part-time users, who in turn demonstrated lower ANL scores than non-hearing aid users. SPIN scores and ANL scores were not strongly correlated. The authors underscored the difference between SPIN and ANL scores, in that SPIN scores indicated the benefit of amplification for speech perception, whereas the ANL scores indicated the difference between successful and unsuccessful hearing aid users. This interpretation of the ANL results was supported by regression analyses that suggest the ANL predicted successful hearing aid outcome with 87% accuracy and unsuccessful hearing aid outcome with 83.6% accuracy. Listeners with ANLs of five or below were expected to be successful users and those with ANLs of fifteen or higher were expected to be unsuccessful.

Some results of this study should be interpreted with caution for a number of reasons. First, the subjects in the study were all experienced hearing aid users—defined as no more than three years of prior hearing aid use. The authors acknowledged that to truly determine predictive value, unaided testing should be conducted prior to hearing aid fitting, but did not specify that it should be done prior to a *first time* hearing aid fitting. Therefore, results obtained in the current study inherently represent previous experience with amplification and previously established hearing aid use patterns.

Second, the authors defined full-time hearing aid users as those subjects who used their hearing aids "whenever they needed them." Many audiologists would question whether this use pattern should truly be considered "full time." Indeed, clinical experience indicates that patients who use their hearing aids most of the day, every day, are the most successful and comfortable with their hearing aids. Most practicing audiologists would consider patients who use their hearing aids "as needed" to be part-time users, unless "as needed" was defined as a significant portion of every day.

Third, the subjects in this study were fitted by their own audiologists in independent clinics, and more importantly, used a wide variety of hearing instruments with different features and signal-processing characteristics. Therefore, their responses to background noise likely varied with signal-processing features that were not controlled variables in this study.

Additional work, with strict controls, should be considered in order to truly investigate the predictive value of the ANL test. Future studies should evaluate new hearing aid users prior to their initial fitting in the unaided condition, then conduct subsequent aided testing periodically during an acclimatization period. Under those circumstances, the unaided ANL could not represent longitudinal effects that have *already occurred* in response to regular hearing aid use. While Nabelek et al. (2004) found that there was no change in ANL or SPIN scores for hearing aid users over the first three months of acclimatization. The participant's previous experience with hearing aids was not controlled; approximately half of the subjects were new users and half had experience with hearing aids. Other studies have shown (Keidser et al., 2008) that acclimatization to hearing aids can extend long beyond the first three months and may continue even beyond one year of use, so the subjective judgments of experienced users could be different from those of inexperienced hearing aid users, even those tested after a few months of use. For this reason, longitudinal study of acceptable noise levels in hearing aid users may be of interest.

One may caution against the use of the ANL as an absolute predictive measure of success with hearing aids. The goal of the audiologist is to rehabilitate hearing impaired individuals, which usually includes fitting them with appropriate hearing aids, counseling, and training. If patients who demonstrate high ANL scores are not expected to be successful with hearing aids, should it follow that these patients be discouraged from trying hearing aids? The authors of this study comment that these patients should be counseled about the "limitations of hearing aids even in quiet listening situations." This conclusion could be detrimental to patient care if hearing aid use is ruled out before the patient is even given a chance to have a trial with them. Regarding subjects with moderate ANL scores, the authors rightly point out the importance of directional microphones and noise reduction, both of which can significantly improve measured ANLs (Pisa et al., 2010). However, these features can be beneficial to all hearing aid users to some degree and would therefore be discussed and recommended regardless of ANL outcome. Perhaps a moderate-to-high ANL score could indicate a need for more aggressive noise reduction at the initial fitting. This is, however, speculation as no work has been published evaluating the effect of changing noise-reduction parameters on ANL scores.

The authors considered their analyses to be an indication that ANL score can predict future success, or at least consistency of use, with hearing aids. However, all of their subjects had previous experience with hearing aids and many had previously established use patterns. Correlation does not implicitly suggest causation; in that regard it should be considered that subjects' use patterns were predictive of their ANL scores, rather than vice versa. Studies of auditory plasticity have shown that auditory experiences, including use of amplification, can affect objective performance and subjective assessments beyond twelve months after the initial fitting (Keidser et al., 2008; Palmer et al., 1998). Therefore, it follows that the subjective acceptance of background noise, as measured by the ANL, can differ between new and experienced users and that changes within the auditory system could result in improved noise tolerance.

Perhaps the most important insight to be gained from the current study is the importance of consistent hearing aid use. The authors of this study found that a consistent pattern of use of hearing aids was highly correlated to low ANL scores. Rather than supporting the clinical use of ANL scores as a predictive tool, these results may instead indicate that consistent hearing aid users have adjusted to amplified sound and have learned to parse complex acoustic information so that they are better able to withstand increasing background noise levels. All hearing aid patients should be counseled about the importance of consistent use, not only to determine the need for future programming modifications, but also to aid their process of acclimatization to amplified sound, including their acceptance of background noise.

References

Bentler, R.A., Niebuhr, J.P., Getta, C.V., & Anderson, C.V. (1993). Longitudinal study of hearing aid effectiveness II: subjective measures. *Journal of Speech and Hearing Research*, 36, 820–831.

Bilger, R.C., Neutzel, J.M., Rabinowitz, W.M., & Rzeczkowski, C. (1984). Standardization of a test of speech perception in noise. *Journal of Speech and Hearing Research*, 27, 32–48.

Humes, L.E., Halling, D., & Coughlin, M. (1996). Reliability and stability of various hearing aid outcome measures in a group of elderly hearing aid wearers. *Journal of Speech, Language and Hearing Research*, 39, 923–935.

Keidser, G., O'Brien, A., Carter, L., McLelland, M., & Yeend, I. (2008). Variation in preferred gain with experience for hearing-aid users. *International Journal of Audiology*, 47:10, 621–635

Nabelek, A.K., Tucker, F.M., Letowski, T.R. (1991). Toleration of background noises: relationship with patterns of hearing aid use by elderly persons. *Journal of Speech and Hearing Research*, 34, 679–685.

Nabelek, A.K., Tampas, J.W., & Burchfield, S.B. (2004). Comparison of speech perception in background noise with acceptance of background in aided and unaided conditions. *Journal of Speech, Language and Hearing Research*, 47, 1001–1011.

Nabelek, A.K., Freyaldenhoven, M.C., Tampas, J.W., Burchfield, S.B., & Muenchen, R.A. (2006). Acceptable noise level as a predictor of hearing aid use. *Journal of the American Academy of Audiology*, 17, 626–639.

Palmer, C.V., Nelson, C.T., & Lindley, G.A. (1998). The functionally and physiologically plastic adult auditory system. *Journal of the Acoustical Society of America*, 103, 1705–1721.

Pisa et al., (2010) missing

Schum, D.J. (1999). Perceived hearing aid benefit in relation to perceived needs. *Journal of the American Academy of Audiology*, 10, 40–45.

Prescribing Compression for Severe Hearing Loss

Souza, P.E., Jenstad, L.M. & Folino, R. (2005). Using multichannel wide-dynamic range compression in severely hearing-impaired listeners: effects on speech recognition and quality. *Ear and Hearing* *26*(2), 120–131.

Most modern hearing aids feature multiple signal-processing channels and wide dynamic range compression (WDRC). For listeners with mild-to-moderate hearing loss, WDRC can offer improvement in speech intelligibility and sound quality in quiet conditions (Souza, 2002). Small benefits of WDRC over linear amplification have also been observed in the presence of background noise (Moore et al., 1999). Most of the research on WDRC has been on individuals with mild or moderate hearing loss. There is considerably less data available on WDRC performance for severely hearing-impaired participants. In fact, many clinical practitioners believe that patients with greater amounts of hearing loss prefer, and benefit most, from linear amplification.

Severe hearing loss is accompanied by reduced frequency selectivity (Faulkner et al., 1990; Rosen et al., 1990) and temporal resolution (Lamore et al., 1990; Nelson & Freyman, 1987). Beyond audibility constraints of hearing loss alone, these impairments further limit ability to identify and discriminate speech cues. Because spectral cues may be limited or unavailable, severely impaired listeners rely on other cues such as variations in the speech amplitude envelope over time (Rosen et al., 1990). For these reasons, many compressor designs are constrained to minimally degrade the speech signal.

One particular concern about WDRC circuitry is that natural variations in speech amplitude may be altered, reducing the availability of amplitude-related cues. The result can be degradation in consonant perception (Souza & Turner, 1998) or overall sentence recognition (Souza & Kitch, 2001b; Stone & Moore, 2003; VanTasell & Trine, 1996). The reduction of amplitude cues, in combination with impaired frequency selectivity, could result in poor performance for severely hearing-impaired individuals using WDRC hearing aids. Indeed, in a study with severely hearing-impaired participants, Souza and Bishop (1999) found smaller improvements in sentence recognition for WDRC amplification than for linear amplification. DeGennaro et al. (1986) also found no advantage with a WDRC system for severely hearing-impaired listeners, despite the fact that the compression system provided improved audibility over the linear system.

In contrast, other studies suggest WDRC benefits for individuals with severe hearing loss. Barker et al. (2001) found that listeners with severe hearing loss preferred single channel WDRC to either output compression limiting or peak clipping. Although user preference is a critical element of a successful hearing aid fitting, it is inarguably more important to ensure adequate speech recognition. While many prescriptive formulas attempt to find a balance, there is no consensus on the most appropriate amplification characteristics to ensure speech recognition and acceptable sound quality for hearing aid users with severe hearing loss.

The purpose of Souza et al.'s study was to examine speech-recognition performance and speech-quality judgments for severely hearing-impaired listeners using four different types of amplification:

1.   Linear with peak clipping
2.   Linear with output compression limiting
3.   Two-channel WDRC
4.   Three-channel WDRC

Thirteen participants with severe, sensorineural hearing loss, most of who had previous hearing aid experience, participated in the study. Seven participants with normal hearing also participated as a control group.

Speech recognition was evaluated using the Nonsense Syllable Test (NST; Resnick et al., 1976). Speech quality judgments were obtained with a paired comparison task, using sentence stimuli from the Connected Speech Test (CST) (Cox et al., 1987). For each stimulus pair, participants heard the same sentence processed in two different amplification conditions and were asked to select the one they preferred. Listeners were specifically instructed to avoid using loudness as a primary criterion for their preference.

Speech materials were processed offline with a master hearing aid simulation of each amplification type described above; signal presentation was at 70dB SPL via an ER-2 insert earphone. Frequency and gain response was determined based on the mean audiometric thresholds of the participant group. The targets themselves represented an average of the NAL-RP (Byrne et al., 1990) and NAL-NL1 (Dillon, 1999) targets for conversational speech.

As would be expected, the speech-recognition scores for normal-hearing participants were high and equivalent for all test conditions. Hearing-impaired participants showed poorer performance for peak clipping and multichannel WDRC conditions than for compression limiting. The only statistically significant difference was between compression limiting and three-channel WDRC.

A more detailed feature analysis was conducted to determine which speech features were poorly transmitted by the three-channel WDRC system, including place, voicing, and manner cues. Place cues were not transmitted effectively by any of the amplification systems, probably because the hearing-impaired participants' had poor frequency resolution and place is mainly transmitted by spectral cues (Rosen, 1992). The two WDRC amplification types preserved voicing information slightly more than compression limiting or peak clipping. Of primary interest was manner of articulation, as it is primarily transmitted via amplitude envelope cues (Rosen, 1992). There is an expectation that amplitude envelope is more distorted by fast-acting WDRC than either peak clipping or output compression limiting.

Detailed analysis of manner transmission revealed that amplification type affected phoneme categories differently. Fricatives were fairly well preserved by compression limiting and WDRC systems but were less well transmitted by the peak clipping system. The two-channel WDRC system transmitted nasality better than peak clipping or compression limiting, but the three-channel WDRC did not preserve nasality cues as well. Because nasality is primarily transmitted via the low-frequency nasal murmur (Kent & Read, 1992) the channel crossover in the three-channel system may have disrupted this cue. Affricates, a combination of a stop and fricative, were best preserved by linear amplification—peak clipping or compression limiting—and were negatively affected by WDRC.

The authors surmised that poor affricate identification may have been related to the time parameters of their WDRC algorithm, causing amplitude decreases after the stop to be perceived as bursts, resulting in misidentification of affricates as stop consonants.

Speech quality ratings by the hearing-impaired participants showed a clear preference for compression limiting. There were significant differences between all comparisons with the exception of peak clipping versus two-channel WDRC. No normal-hearing participants preferred peak clipping and their preferences were equally divided among the other options.

Overall, the results of this study indicate poorer speech recognition and sound quality for WDRC compared to compression limiting. Previous work indicated that compression benefit decreased as pure-tone thresholds exceeded 70dBHL (Goedegebure et al., 2001; Souza & Bishop, 1999), and Boothroyd et al. (1988) concluded that amplitude distortions were responsible for poor performance with a two-channel, fast-acting compression system. These assumptions may be supported by the current work, in which stop-affricate confusions were considered to be related to misinterpretation of amplitude cues.

Though compression limiting yielded better speech recognition and sound quality in the present study, the authors caution that this should not preclude the use of WDRC systems with severely hearing-impaired individuals. For example, they acknowledge that the short release times they used may have affected amplitude envelope cues. Because of the wide variability in parameters among current WDRC circuits, there will be variability in their effect on speech envelope cues as well. Preservation of speech amplitude envelope is likely to aid word recognition and clarity for users with severe loss and may be achieved with longer attack and release times (Kuk & Ludvigsen, 2000). Alternatively, shorter attack and release times, combined with lower compression ratios may also reduce the detrimental effects of WDRC on amplitude cues.

Souza and her colleagues point out that the frequency responses they used in the study may have deviated from individually prescribed targets, compromising performance. They applied their prescriptive formula to the average hearing loss of their participant population, which may not have provided as close a match to target as would be achieved in a typical hearing aid fitting. In general, their frequency responses over amplified slightly in the low frequencies and under amplified in the high frequencies. They used a fixed compression ratio of 3:1 across channels for all participants, higher than those prescribed by prescriptive procedures for clinical use; additionally, the use of one ratio across channels does not represent the typical prescription. These deviations from clinically prescribed settings may have affected the speech-recognition scores and sound-quality preferences in their study.

The hearing-impaired individuals in this study clearly preferred the sound quality of compression limiting over the peak clipping or WDRC options. The improved audibility of high-frequency information with the WDRC systems may have adversely affected sound quality for some listeners, especially those not accustomed to high-frequency emphasis. Clinicians are well aware that when individuals wear new hearing instruments with better high-frequency amplification, their initial description of the sound quality is "tinny" or "metallic." It can take days or even weeks for them to adjust to the new frequency response, before the increased high-frequency audibility is received favorably.

Similarly, many experienced hearing aid users react negatively to WDRC because it sounds softer than the linear amplification they have become accustomed to. Participants in this study were specifically instructed to disregard loudness in their sound-quality judgments, but the relative decrease in loudness in the WDRC trials may have affected their preferences anyway. For experienced users of powerful hearing instruments, a decrease in loudness can be interpreted initially as a decrease in performance, despite accompanying improvements in speech-recognition ability.

While educative, these findings are based on atypical compressor parameters; thus, applying them directly to clinical decision making should be done with caution. Beyond the considerations of gain and time constants, sound-quality judgments based on everyday performance might be very different, especially after use over a longer period of time. Hearing aid users are known to experience a period of acclimatization to new hearing aids, the duration of which varies depending on their degree of loss and prior user of amplification (Keidser et al., 2009). It is reasonable to assume that a WDRC circuit providing superior high-frequency audibility, though initially perceived as "too soft" or "too tinny," would sound more natural and clear after an extended period of consistent use.

Souza and her colleagues address an important issue for clinicians fitting individuals with severe-to-profound hearing loss.

As they acknowledge, more study is needed regarding specific WDRC parameters and how they affect speech recognition and preferences in the severely hearing-impaired population. Of particular interest is how WDRC parameters perform over time in everyday listening situations, outside of the laboratory. Findings, such as these, illustrate how objective and subjective measures reveal discrete aspects of a patient's experience, proving information that helps clinicians determine the most appropriate starting point for these individuals.

References

Barker, C., Dillon, H. & Newall, P. (2001). Fitting low ratio compression to people with severe and profound hearing losses. *Ear and Hearing*, 22, 130–141.

Byrne, D., Parkinson, A. & Newall, P. (1990). Hearing aid gain and frequency response requirements for the severely-profoundly hearing impaired. *Ear and Hearing*, 11, 40–49.

Cox, R.M., Alexander, G.C. & Gilmore, C. (1987). Development of the Connected Speech Test (CST). *Ear and Hearing* 8 (Suppl.), 119S–126S.

DeGennaro, S., Braida, L.D. & Durlach, N.I. (1986). Multichannel syllabic compression for severely impaired listeners. *Journal of Rehabilitation Research and Development*, 23, 17–24.

Dillon, H. (1999). NAL-NL1: A new prescriptive fitting procedure for non-linear hearing aids. *Hearing Journal*, 52, 10–16.

Faulkner, A., Rosen, S. & Moore, B.C.J. (1990). Residual frequency selectivity in the profoundly hearing-impaired listener. *British Journal of Audiology*, 24, 381–392.

Humes, L.E., Christensen, L., Thomas, T., Bess, F., Hedley-Williams, A. & Bentler, R. (1999). A comparison of the aided performance and benefit provided by a linear and a two-channel wide dynamic range compression hearing aid. *Journal of Speech, Language and Hearing Research*, 42, 65–79.

Keidser G., O'Brien, A., Carter, L., McLelland, M. & Yeend, I. (2008). Variation in preferred gain with experience for hearing-aid users. *International Journal of Audiology*, 47(10), 621–35.

Kent, R.D. & Read, C. (1992). *The acoustic analysis of speech*. San Diego: Singular Publishing Group.

Kuk, F. & Ludvigsen, C. (2000). Hearing aid design and fitting solutions for persons with severe-to-profound losses. *Hearing Journal*, 53, 29–37.

Lamore, P.J., Verweij, C. & Brocaar, M.P. (1990). Residual hearing capacity of severely hearing-impaired participants. *Acta Otolaryngologica Supplement*, 469, 7–15.

Moore, B.C.J., Peters, R.W. & Stone, M.A. (1999). Benefits of linear amplification and multi-channel compression for speech comprehension in backgrounds with spectral and temporal dips. *Journal of the Acoustical Society of America*, 105, 400–411.

Nelson, D.A. & Freyman, R.L. (1987). Temporal resolution in sensorineural hearing-impaired listeners. *Journal of the Acoustical Society of America*, 81, 709–720.

Resnick, S.B., Dubno, J.R., Hoffnung, S. & Levitt, H. (1976). Phoneme errors on a nonsense syllable test. *Journal of the Acoustical Society of America*, 58 (Suppl. 1), 114.

Rosen, S. (1992). Temporal information in speech: Acoustic, auditory and linguistic aspects.*Philosophical transactions of the Royal Society of London Series B: Biological Sciences* 336, 367–373.

Rosen, S., Faulkner, A. & Smith, D.A. (1990). The psychoacoustics of profound hearing impairment. *Acta Otolaryngologica Supplement*, 469, 16–22.

Souza, P.E. (2002). Effects of compression on speech acoustics, intelligibility and sound quality. *Trends in Amplification*, 6, 131–165.

Souza, P.E. & Bishop, R. (1999). Improving speech audibility with wide dynamic range compression in listeners with severe sensorineural hearing loss. *Ear and Hearing*, 20, 461–470.

Souza, P.E., Jenstad, L.M. & Folino, R. (2005). Using multichannel wide-dynamic range compression in severely hearing-impaired listeners: effects on speech recognition and quality. *Ear and Hearing,* 26(2), 120–131.

Souza, P.E. & Kitch, V.J. (2001b). The contribution of amplitude envelope cues to sentence identification in young and aged listeners. *Ear and Hearing,* 22, 112–119.

Souza, P.E. & Turner, C. W. (1998). Multichannel compression, temporal cues and audibility. *Journal of Speech and Hearing Research,* 41, 315–326.

Stone, M.A. & Moore, B.C.J. (2003). Effect of the speed of a single-channel dynamic range compressor on intelligibility in a competing speech task. *Journal of the Acoustical Society of America,* 114, 1023–1034.

VanTasell, D. J. & Trine, T.D. (1996). Effect of single-band syllabic amplitude compression on temporal speech information in nonsense syllables and in sentences. *Journal of Speech and Hearing Research,* 39, 912–922.

Transitioning the Patient with Severe Hearing Loss to New Hearing Aids

Convery, E., & Keidser, G. (2011). Transitioning hearing aid users with severe and profound loss to a new gain/frequency response: benefit, perception and acceptance. *Journal of the American Academy of Audiology, 22*, 168–180.

Many individuals with severe-to-profound hearing loss are full-time, long-term hearing aid users. Because they rely heavily on their hearing aids for everyday communication, they are often reluctant to try new technology. It is common to see patients with severe hearing loss keep a set of hearing aids longer than those with mild-to-moderate losses. These older hearing aids offered less effective feedback suppression and a narrower frequency range than those available today now. The result was that many severely impaired hearing aid users were fitted with inadequate high-frequency gain and compensatory increases in low-mid frequency amplification. Having adapted to this frequency response, they may reject new hearing aids with increased high-frequency gain, stating that they sound too tinny or unnatural. Similarly, those who have adjusted to linear amplification may reject wide dynamic range compression (WDRC) as too soft, even though it the strategy may provide some benefits when compared to their linear hearing aids.

Convery and Keidser evaluated a method to gradually transition experienced, severely impaired hearing aid users into new amplification characteristics. They measured subjective and objective outcomes as subjects took incremental steps toward a more appropriate frequency response. Twenty-three experienced, adult hearing aid users participated in the study. Participation was limited to subjects whose current gain and frequency response differed significantly from targets based on NAL-RP, a modification of the NAL formula for severe-to-profound hearing losses (Byrne et al.

1991). Most subjects' own instruments had more gain at 250–2000Hz and less gain at 6–8 kHz compared to NAL-RP targets, so the experimental transition involved adapting to less low- and mid-frequency gain and more high-frequency gain.

Subjects in the experimental group were fitted bilaterally with WDRC behind-the-ear hearing instruments. Directional microphones, noise reduction, and automatic features were turned off and volume controls were activated with an 8dB range. The hearing aids had two programs: the first, called the "mimic" program, had a gain/frequency response adjusted to match the subject's current hearing aids; the second program was set to NAL-RP targets. MPO was the same for mimic and NAL-RP programs. The programs were not manually accessible for the user; they were only adjusted by the experimenters at test sessions.

Four incremental programs were created for each participant in the experimental group. Each step was approximately a 25% progression from their mimic program-frequency response to the NAL-RP prescribed response. At three-week intervals, they were switched to the next incremental program, approaching NAL-RP settings as the experiment progressed. The programs in the control group's hearing aids remained consistent for the duration of the study.

All subjects attended eight sessions. At the initial session, subjects' own instruments were measured in a 2cc coupler and RECD measurements were obtained with their own earmolds. The experimental hearing aids were fitted at the next session and subjects returned for follow-up sessions at one week post fitting and three week intervals thereafter until fifteen weeks post fitting.

Subjects evaluated the mimic and NAL-RP programs in paired comparisons at one week and fifteen weeks post fitting. The task used live dialogues with female talkers in four everyday environments: café, office, reverberant stairwell, and outdoors with traffic noise in the background. Hearing aid settings were switched from mimic to NAL-RP with a remote control, without audible program-change beeps, so subjects were unaware of their current program. They were asked to indicate their preference for one program over the other on a four-point scale: no difference, slightly better, moderately better, or much better.

Speech discrimination was evaluated with the Beautifully Efficient Speech Test (BEST; Schmitt, 2004), which measured the aided SRT for sentence stimuli. Loudness scaling was then conducted to determine the most comfortable loudness level and range (MCL/R). Finally, subjects responded to a questionnaire concerning overall loudness comfort, speech intelligibility, sound quality, use of the volume control, use of their own hearing aids, and perceived changes in audibility and comfort. Speech discrimination, loudness scaling, and questionnaire administration took place for all participants at three-week intervals, starting at the three-week post-fitting session.

One goal of the study was to determine if there would be a change in speech discrimination over time or a difference between the experimental and control groups. Analysis of BEST SRT scores yielded no significant difference between the experimental and control groups, nor was there a significant change in SRT over time. There was a significant interaction between these variables, indicating that the experimental group demonstrated slightly poorer SRT scores over time, whereas the control group's SRTs improved slightly over time.

Subjects rated perceptual disturbance, or how much the hearing aid settings in the current test period differed from the previous period and how disturbing the difference was. There was no significant effect for the experimental or control groups, but there was a tendency for reports of perceptual disturbance over time to decrease for the control group and increase for the experimental group. The mimic programs for the control group were consistent, so control subjects likely became acclimated over time. The experimental group, however, had incremental changes to their mimic program at each session, so it is not surprising that they reported more perceptual disturbance. This was only a slight trend, however, indicating that even the experimental group experienced relatively little disturbance as their hearing aids approached NAL-RP targets.

Analysis of the paired-comparison responses indicated a significant overall preference for the mimic program over the NAL-RP program. There was an interaction between environment and listening program, showing a strong preference for the mimic program in office and outdoor environments and somewhat less of a preference in the café and stairwell environments. When asked about their criteria for the comparisons, subjects most commonly cited speech clarity, loudness, comfort, and naturalness, regardless of whether mimic fit or NAL-RP was preferred. There was no significant effect of time on program preference, but there was a slight increase in the control group's preference for mimic at the end of the study, whereas the experimental group shifted slightly toward NAL-RP, away from mimic.

Over the course of the study, Convery and Keidser's subjects demonstrated acceptance of new frequency responses with less low- to mid-frequency gain and more high-frequency gain than their current hearing aids. No significant differences were noted between experimental and control groups for loudness, sound quality, voice quality, intelligibility, or overall performance, nor did these variables change significantly over time. Though all subjects preferred the mimic program overall, there was a trend for the experimental group to shift slightly toward a preference for the NAL-RP settings, whereas the control group did not. This indicates that the experimental subjects had begun to acclimate to the new, more appropriate frequency response. Acclimatization might have continued to progress had the study examined performance over a longer period of time. Prior research indicates that acclimatization to new hearing aids can progress over the course of several months and individuals with moderate and severe losses may require more time to adjust than individuals with milder losses (Keidser et al., 2008).

Reports of perceptual disturbance increased as incremental programs approached NAL-RP settings. This may not be surprising to clinicians, as hearing aid patients often require a period of acclimatization even after relatively minor changes to their hearing aid settings. Furthermore, clinical observation supports the suggestion that individuals with severe hearing loss may be even more sensitive to small changes in their frequency response. Allowing more than three weeks between program changes may result in less perceptual disturbance and easier transition to the new frequency response. Clinically, perceptual disturbance with a new frequency response can also be mitigated by counseling and encouraging patients that they will feel more comfortable with the new hearing aids as they progress through their trial periods. It might also be helpful to extend the trial period (which is usually thirty to forty-five days) for individuals with severe-to-profound hearing losses, to accommodate an extended acclimatization period.

Individuals with severe-to-profound hearing loss often hesitate to try new hearing aids. Similarly, audiologists may be reluctant to recommend new instruments with WDRC or advanced features for fear that they will be summarily rejected. Convery and Keidser's results support a process for transitioning experienced hearing aid users into new technology and suggest an alternative for clinicians who might otherwise hesitate to attempt departures from a patient's current frequency response.

Because this was a double-blind study, the research audiologists were unable to counsel subjects as they would in a typical clinical situation. The authors note that counseling during transition is of particular importance for severely impaired hearing aid users, to ensure realistic expectations and acceptance of the new technology. Though the initial fitting may approximate the client's old frequency response, follow-up visits at regular intervals should slowly implement a more desirable frequency response. Periodically, speech discrimination and subjective responses should be evaluated and the transition should be stopped or slowed if decreases in intelligibility or perceptual disturbances are noted.

In addition to changes in the frequency response, switching to new hearing aid technology usually means the availability of unfamiliar features such as directional microphones, noise reduction, and many wireless features. Special features such as these can be introduced after the client acclimates to the new frequency response, or they can be relegated to alternate programs to be used on an experimental basis by the client. For instance, automatic directional microphones are sometimes not well received by individuals who have years of experience with omnidirectional hearing aids. By offering directionality in an alternate program, the individual can test it out as needed and may be less likely to reject the feature or the hearing aids. It is critical to discuss proper use of the programs and to set up realistic expectations. Because variable factors such as frequency resolution and sensitivity to incremental amplification changes may affect performance and acceptance, the transition period should be tailored to the needs of the individual and monitored closely with regular follow-up appointments.

References

Baer, T., Moore, B.C.J. & Kluk, K. (2002). Effects of low pass filtering on the intelligibility of speech in noise for people with and without dead regions at high frequencies. *Journal of the Acoustical Society of America, 112*, 1133–1144.

Barker, C., Dillon, H. & Newall, P. (2001). Fitting low ratio compression to people with severe and profound hearing losses. *Ear and Hearing, 22*, 130–141.

Byrne, D., Parkinson, A. & Newall, P. (1991). Modified hearing aid selection procedures for severe/profound hearing losses. In:

Studebaker, G.A., Bess, F.H., Beck, L. eds. *The Vanderbilt Hearing Aid Report II*. Parkton, MD: York Press, 295–300.

Ching, T.Y.C., Dillon, H., Lockhart, F., vanWanrooy, E. & Carter, L. (2005). Are hearing thresholds enough for prescribing hearing aids? Poster presented at the 17th Annual American Academy of Audiology Convention and Exposition, Washington, DC.

Convery, E. & Keidser, G. (2011). Transitioning hearing aid users with severe and profound loss to a new gain/frequency response: benefit, perception and acceptance. *Journal of the American Academy of Audiology, 22*, 168–180.

Flynn, M.C., Davis, P.B. & Pogash, R. (2004). Multiple-channel non-linear power hearing instruments for children with severe hearing impairment: long-term follow-up. *International Journal of Audiology, 43*, 479–485.

Keidser, G., Hartley, D. & Carter, L. (2008). Long-term usage of modern signal processing by listeners with severe or profound hearing loss: a retrospective survey. *American Journal of Audiology, 17*, 136–146.

Keidser, G., O'Brien, A., Carter, L., McLelland, M., and Yeend, I. (2008) Variation in preferred gain with experience for hearing-aid users. *International Journal of Audiology, 47*(10), 621–635.

Kuhnel, V., Margolf-Hackl, S. & Kiessling, J. (2001). Multi-microphone technology for severe to profound hearing loss. Scandanavian Audiology 30 (Suppl. 52), 65–68.

Moore, B.C.J. (2001). Dead regions in the cochlea: diagnosis, perceptual consequences and implications for the fitting of hearing aids. *Trends in Amplification, 5*, 1–34.

Moore, B.C.J., Killen, T. & Munro, K.J. (2003). Application of the TEN test to hearing-impaired teenagers with severe-to-profound hearing loss. *International Journal of Audiology, 42*, 465–474.

Schmitt, N. (2004). *A New Speech Test (BEST Test). Practical Training Report*. Sydney: National Acoustic Laboratories.

Vickers, D.A., Moore, B.C.J. & Baer, T. (2001). Effect of low-pass filtering on the intelligibility of speech in quiet for people with and without dead regions at high frequencies. *Journal of the Acoustical Society of America, 110*, 1164–1175.

Do Patients with Severe Hearing Loss Benefit from Directional Microphones?

Ricketts, T.A., & Hornsby, B.W.Y. (2006). Directional hearing aid benefit in listeners with severe hearing loss. *International Journal of Audiology, 45*, 190–197.

The benefit of directional microphones for speech recognition in noise is well established for individuals with mild-to-moderate hearing loss (Madison & Hawkins, 1983; Killion et al., 1998; Ricketts 2000a; Ricketts & Henry, 2002). The potential benefit of directional microphones for severely hearing-impaired individuals is less understood and few studies have examined directional benefit when hearing loss is greater than 65dB.

Killion and Christensen (1998) proposed that listeners with severe-to-profound hearing loss may experience reduced directional benefit because they are less able to make use of speech information across frequencies. Ricketts, Henry and Hornsby confirmed this hypothesis in a 2005 study. They found an approximately 7% increase in speech-recognition score per 1dB increase in directivity for listeners with moderate hearing loss, whereas listeners with severe loss achieved only an approximately 3.5% increase per 1dB increase in directivity.

In their 2005 study, Ricketts and Hornsby used individually determined SNRs and auditory-visual stimuli that allowed testing at poorer SNRs without floor effects. The authors point out that visual cues usually offer a greater benefit at poor SNRs, especially for sentence materials (Erber, 1969; Sumby & Pollack, 1954; MacLeod & Summerfield, 1987). Individuals rely more on visual cues in poorer SNRs, visual information that provides complementary, non-redundant cues is most beneficial (Grant, 1998; Walden et al., 1993).

The purpose of their study was to examine potential directional benefit for severely hearing-impaired listeners at multiple SNRs in auditory-only and auditory-visual conditions. Directional and omnidirectional performance in quiet conditions were also tested to rule out performance differences between microphone modes that could be attributed to reduction of environmental noise by the directional microphone. Finally, it was determined if performance in quiet conditions would significantly exceed performance in highly positive SNRs. Though significant improvement in SNRs more favorable than +15 dB is usually not expected, some research suggests that hearing-impaired individuals may experience additional benefit from more favorable SNRs (Studebaker et al., 1999).

Twenty adult participants with severe-to-profound sensorineural hearing loss participated in the study. All participants used oral communication, had at least nine years of experience with hearing aids, and had pure-tone average hearing thresholds greater than 65dB. Participants were fitted with power behind-the-ear hearing aids with full-shell, unvented earmolds. Digital noise reduction and feedback management was turned off. The directional program was equalized, so that gain matched the omnidirectional mode as closely as possible.

The Audio/Visual Connected Speech Test (CST; Cox et al., 1987), a speech-recognition test with paired passages of connected speech, was presented to listeners on DVD. Speech was presented at a 0-degree azimuth angle and uncorrelated competing noise was presented via five loudspeakers surrounding the listener. Testing took place in a sound booth with reflective panels to approximate common levels of reverberation in everyday situations.

Baseline SNRs were obtained for each subject in auditory-only and auditory-visual conditions, at a level that was near, but not reaching floor performance. Speech-recognition testing was conducted for omnidirectional and directional conditions at baseline SNR, baseline + 4dB and baseline + 8dB. Presentation SNRs ranged from 0dB to +24dB for auditory-only conditions and from -6dB to +18dB for auditory-visual conditions. Listeners were tested with auditory-only stimuli in quiet conditions, for omnidirectional and directional modes. Testing in quiet was not performed with auditory-visual stimuli, as performance was expected to approach ceiling performance levels.

The multiple SNR levels were achieved with two different methodologies. Half of the participants listened to a fixed noise level of 60dB SPL and speech levels were varied to achieve the desired SNRs. The remaining participants listened to a fixed speech level of 67dB SPL and the noise levels were adjusted to reach the desired SNR levels. Data analysis revealed no significant differences between these two test methodologies for any of the variables, so their data was pooled for subsequent analyses.

The results showed significant main effects for microphone mode (directional versus omni), SNR, and presentation condition (auditory-only versus auditory-visual). There were significant interactions between microphone mode and SNR, as well as between presentation condition and SNR. Each increase in SNR resulted in significantly better performance for both omnidirectional and directional modes. Performance in directional mode was significantly better than omnidirectional for all SNR levels. The authors pointed out that auditory-visual performance at all three SNRs was always better than auditory-only, despite the fact that the absolute SNRs for auditory-visual conditions were lower than the equivalent auditory-only conditions, by an average of 5dB. The authors interpreted this finding as strong support for the benefit of visual cues for speech recognition in adverse conditions.

When the effects of directionality and SNR were analyzed separately for auditory-only and auditory-visual conditions, they found that directional performance was significantly better than omnidirectional performance for all auditory-visual conditions. In auditory-only conditions, directionality only had a significant effect at the baseline SNR, but not in the baseline +4dB, baseline +8dB, or quiet conditions.

Perhaps not surprisingly, Ricketts and his colleagues found that the addition of visual cues offered their severely impaired listeners a significant advantage for understanding connected speech. When they compared the auditory-only and auditory-visual scores at equivalent SNR levels, they determined that participants achieved an average improvement of 22% with the availability of visual cues. This finding is in agreement with previous research that found a visual advantage of 24% for listeners with moderate hearing loss (Henry & Ricketts, 2003).

Also not surprisingly, performance improved with increases in signal-to-noise ratio. For the auditory-only condition, they found an average improvement of 1.6% per dB and 2.7% per dB for the omnidirectional and directional modes, respectively. For the auditory-visual condition, there was an improvement of 3.7% per dB for omnidirectional mode and 3.1% per dB for directional mode. Furthermore, they found an additional performance increase of 8% for directional mode and 12% for omnidirectional mode when participants were tested in quiet conditions. This was somewhat surprising given previous research based on the articulation index (AI) that suggested maximal performance could be expected at SNRs of approximately +15dB. The absolute SNR for the baseline +8dB condition was 14.7dB, so further improvements in quiet conditions support the suggestion that hearing-impaired listeners experience increased improvement for SNRs up to +20dB (Studebaker et al., 1999; Sherbecoe & Studebaker, 2002).

The benefit of visual cues was not specifically addressed by this study because it did not compare auditory-only and auditory-visual performance at the same SNR levels. However, the discovery that visual cues improved performance even when the SNRs were approximately 5dB poorer was strong support for the benefit of visual information for speech recognition in noisy environments. This underscores the recommendation that severely hearing-impaired listeners should always be counseled to take advantage of visual cues whenever possible, especially in adverse listening conditions. Although visual cues cannot completely counterbalance the auditory cues lost to hearing loss and competing noise, they supply additional information that can help the listener identify or differentiate phonemes, especially in connected speech containing semantic and syntactic context. In conversational situations, visual cues include not just lip-reading but also the speaker's gestures, expressions, and body language. All of these cues can aid speech recognition, so hearing-impaired individuals as well as their family members should be trained in strategies to maximize the availability of visual information.

Ricketts and Hornsby's study supports the potential benefit of directional microphones for individuals with severe hearing loss. Many hearing aid users with severe-to-profound loss have become accustomed to the use of omnidirectional microphones and may be resistant to directional microphones, especially automatic directionality, if it is in the primary program of their hearing instruments. One strategy for addressing these cases is to program the hearing aid's primary memory as full-time omnidirectional while programming a second, manually accessed memory with a full-time directional microphone. This way the listener is able to choose when and how they use their directional program and may be less likely to experience unexpected and potentially disconcerting changes in perceived loudness and sound quality.

In addition to providing evidence for the benefit of visual cues and directionality, the findings of this study can be extrapolated to support the use of FM and wireless accessories. The fact that performance in quiet conditions was still significantly better than the next most favorable SNR (14.7dB) shows that improving SNR as much as possible provides demonstrable advantages for listeners with severe hearing loss. Even for individuals who do well with their hearing instruments overall, wireless accessories that stream audio directly to the listeners hearing instruments may further improve understanding. These accessories improve SNR by reducing the effect of room acoustics and reverberation, as well as reducing the effect of competing noise and distance between the sound source and the listener. Most modern hearing instruments are compatible with wireless accessories so hearing aid evaluations should always include discussion of their potential benefits. These devices work with a wide range of hearing aid styles, do not require the use of an adapter or receiver boot, and are much less expensive than an FM system.

Ricketts and Hornsby's study underscores the importance of visual information and directionality for speech recognition in noisy environments and illuminates ways in which clinicians can help patients with severe-to-profound loss achieve improved communication in everyday circumstances. Modern technologies such as directional processing and wireless audio streaming accessories can be effective tools for improving SNRs in everyday situations that may otherwise challenge or overwhelm the listener with severe-to-profound hearing loss.

References

Erber, N.P. (1969). Interaction of audition and vision in the reception of oral speech stimuli. *Journal of Speech and Hearing Research*, 12, 423–425.

Grant, K.W., Walden, B.E. & Seitz, P.F. (1998). Auditory-visual speech recognition by hearing-impaired subjects: consonant recognition, sentence recognition and auditory-visual integration. *Journal of the Acoustical Society of America,* 103, 2677–2690.

Henry, P. & Ricketts, T. A. (2003). The effect of head angle on auditory and visual input for directional and omnidirectional hearing aids. *American Journal of Audiology,* 12(1), 41–51.

Killion, M. C., Schulien, R., Christensen, L., Fabry, D. & Revit, L. (1998). Real world performance of an ITE directional microphone. *Hearing Journal,* 51(4), 24–38.

Killion, M.C. & Christensen, L. (1998). The case of the missing dots: AI and SNR loss. *Hearing Journal,* 51(5), 32–47.

MacLeod, A. & Summerfield, Q. (1987). Quantifying the contribution of vision to speech perception in noise. *British Journal of Audiology,* 21, 131–141.

Madison, T.K. & Hawkins, D.B. (1983). The signal-to-noise ratio advantage of directional microphones. *Hearing Instruments,* 34(2), 18, 49.

Pavlovic, C. (1984). Use of the articulation index for assessing residual auditory function in listeners with sensorineural hearing impairment. *Journal of the Acoustical Society of America,* 75, 1253–1258.

Pavlovic, C., Studebaker, G. & Scherbecoe, R. (1986). An articulation index based procedure for predicting the speech recognition performance of hearing-impaired individuals. *Journal of the Acoustical Society of America,* 80(1), 50–57.

Ricketts, T.A. (2000a). Impact of noise source configuration on dire3ctional hearing aid benefit and performance. *Ear and Hearing,* 21(3), 194–205.

Ricketts, T.A. & Henry, P. (2002). Evaluation of an adaptive directional microphone hearing aid. *International Journal of Audiology,* 41(2), 100–112.

Ricketts, T., Henry, P. & Hornsby, B. (2005). Application of frequency importance functions to directivity for prediction of benefit in uniform fields. *Ear & Hearing*, 26(5), 473–86.

Studebaker, G., Sherbecoe, R., McDaniel, D. & Gwaltney, C. (1999). Monosyllabic word recognition at higher-than-normal speech and noise levels. *Journal of the Acoustical Society of America*, 105(4), 2431–2444.

Sherbecoe, R.L., Studebaker, G.A. (2002). Audibility-index functions for the Connected Speech Test. *Ear & Hearing*, 23(5), 385–398.

Sumby, W.H. & Pollack, I. (1954). Visual contribution to speech intelligibility in noise. *Journal of the Acoustical Society of America*, 26, 212–215.

Walden, B.E., Busacco, D.A. & Montgomery, A.A. (1993). Benefit from visual cues in auditory-visual speech recognition by middle-aged and elderly persons. *Journal of Speech and Hearing Research*, 36, 431–436.

Impact of Classroom Noise on Children's Listening

Howard, C. S., Munro, K., & Plack, C. J. (2010). Listening effort at signal-to-noise ratios that are typical of the school classroom. *International Journal of Audiology*, *49*, 928–932.

Everyday activities often require attention to more than one concurrent task. The ability to do this successfully depends on a number of factors, including distractions, the difficulty of the tasks, and the perceived importance of the tasks. In a classroom, children regularly have to attend to multiple tasks at the same time. For instance, they may be taking notes and reading information on a board or a computer screen, while also listening to the teacher and comments or questions from other students. To complicate matters, these tasks are often carried out in the presence of varying levels of background noise.

Classroom noise has a detrimental effect on learning (Shield & Dockrell, 2003). Completing more than one task at a time in a noisy place may adversely affect learning because it requires greater listening effort on behalf of the student. In other words, in the presence of background noise and when attending to multiple tasks, greater cognitive resources must be dedicated to understanding speech. This means that performance on one or more of the tasks, including comprehension of the spoken lesson, can deteriorate. Classroom signal-to-noise ratios (SNRs) have been measured in the range of -7dB to +5dB (Arnold & Canning, 1999; Crandell & Smaldino, 1995, 2000). Low SNRs are known to have a particularly detrimental effect on speech perception for hearing-impaired listeners, especially children (Blandy & Lutman, 2005; Jamieson et al. 2004). Therefore, the effect of SNR on listening effort and classroom multi-tasking are of special concern for hearing-impaired students.

Listening effort can be measured in adults with self-report ratings, in children it is usually measured with dual-task paradigms. Hicks and Tharpe (2002) compared the performance of children with mild hearing loss to that of normal hearing children in a dual-task study. The primary task was word recognition at 70dB SPL in quiet and in multi-talker babble at SNRs of +10dB to +20dB. The secondary task measured visual reaction time to randomly presented lights. The authors found that reaction time was longer for the hearing-impaired children than for the normal-hearing children, suggesting that the hearing-impaired children required more listening effort, therefore devoting fewer cognitive resources to the secondary task. Interestingly, there was no significant effect of SNR on listening effort.

McFadden and Pittman (2008) conducted a dual-task experiment with hearing-impaired and normal-hearing eight- to twelve-year olds. The primary task was to categorize words, presented in quiet and in noise at SNRs or 0dB and +6dB. The secondary task involved completion of a dot-to-dot puzzle. Performance on the secondary task decreased when both tasks were performed together, though there was no significant effect of hearing loss or SNR.

The authors of the current study surmised that the SNRs used in previous studies were too favorable and did not represent typical classroom SNRs. SNR may indeed have a detrimental effect on multi-tasking, but the SNRs used in previous experiments might not have been sufficiently challenging to yield an effect. The study summarized in this blog post aimed to measure listening effort in a dual task paradigm using SNRs that were more typical of classroom environments. The authors hypothesized that as SNR decreased, listening effort would increase, yielding poorer performance on the secondary task. Thirty-one normal-hearing children, age nine to twelve years, participated in the study. None of the subjects had any history of hearing loss or communication or learning disabilities.

The primary task was a word-recognition test using consonant-vowel-consonant monosyllables (Boothroyd, 1968). Words were spoken by a male speaker at 65dB SPL and presented binaurally via headphones. Each set of words was mixed with multi-speaker babble that had been recorded from children's background chatter (Hamilton, 2008) that the authors deemed most similar to typical classroom background noise. The level of babble was adjusted to create four SNR conditions: quiet,

+4dB, 0dB and -4dB. The secondary task involved rehearsing sets of five visually presented digits and reciting them at a later time. Each task was presented alone and together in a dual-task condition. In the dual-task condition, the string of five digits was presented for three seconds. Then, a set of five words was presented and scored before the subjects were asked to recall the rehearsed digits.

For performance on the primary task, the authors found a significant effect of SNR and task combination, as well as a significant interaction between SNR and task combination. In other words, performance on the word-recognition task deteriorated when combined with the digit-recall task, and also deteriorated with decreasing SNR. Even more deterioration in performance was noted in the dual-task condition when SNR decreased.

For performance on the secondary task, there was less of an effect of SNR in the single-task condition, suggesting that subjects were able to ignore the background noise successfully as they completed the visual-recall task. For the dual-task condition, digit-recall performance decreased significantly, especially for lower SNRs. There was a strong, significant interaction between SNR and task combination, showing that performance decreased more substantially in the dual-task condition when SNRs were lower.

As expected, mean performance on the word-recognition test decreased with lower SNRs. In general, the dual-task condition yielded similar word-recognition performance, but for lower SNRs, performance on the secondary task deteriorated, supporting the hypothesis that increased listening effort was required for multi-tasking in the presence of increasing background noise. This is consistent with results found for adult subjects, in which decreased performance on a secondary reaction-time task was found when noise was added to the primary listening task (Downs & Crum, 1978).

All test stimuli used in the current experiment were presented under headphones. This test condition removes acoustic cues that would be experienced in the classroom where there is spatial separation between primary and secondary speech stimuli. Future work with hearing-impaired children listening in a spatially distributed sound field would more closely approximate a classroom environment. Additionally, the use of hearing aids with directional microphones and FM systems may reduce some deleterious effects of SNR. Further research is needed to illuminate the interactions among these variables.

The current study showed a clear deterioration in secondary task performance as the SNR decreased, suggesting that increased listening effort was required in conditions with poorer SNRs. This has important implications for classroom environments, in which children are regularly required to listen and perform secondary tasks such as taking notes and reading visual materials. As classroom background noise increases, children are likely to have fewer cognitive resources available to attend to the spoken lesson, take notes and participate in discussions. Decreases in classroom noise levels may be achieved in many ways, including classroom architecture and design, the use of acoustic damping and noise-reduction materials, and organization of students' workstations.

These observations have particular importance for hearing-impaired students, who are more likely to suffer deleterious effects of classroom background noise. Although this study did not include hearing-impaired listeners, the findings support the continued recommendation of preferential seating, FM systems, and other efforts to improve signal-to-noise ratio in the classroom.

References

Arnold, P., & Canning, D. (1999). Does classroom amplification aid comprehension? *British Journal of Audiology*, 33, 171–178.

Blandy, S., & Lutman, M. (2005). Hearing threshold levels and speech recognition in noise in 7-year-olds. *International Journal of Audiology*, 44, 435–443.

Boothroyd, A. (1968). Developments in speech audiometry. *British Journal of Audiology*, 2, 3–10.

Crandell, C.C. & Smaldino, J.J. (1995). Speech perception in the classroom. In C. Crandell, J. Smaldino & C. Flexer (Eds.), *Sound-field FM Amplification: Theory and Practical Applications*. San Diego: Singular Publication Group.

Crandell, C.C., & Smaldino, J.J. (2000). Classroom acoustics for children with normal hearing and with hearing impairment. *Language, Speech and Hearing Services in Schools*, 31, 362–370.

Downs, D.W., & Crum, M.A. (1978). Processing demands during auditory learning under degraded listening conditions. *Journal of Speech and Hearing Research*, 21, 702–714.

Hamilton, G. (2008). *Compressed Babble for Speech-in-Noise Testing*. Available from: The Ewing Foundation in association with the University of Manchester. http://www.ewing-foundation.org.uk/.

Hicks, C.B., & Tharpe, A.M. (2002). Listening effort and fatigue in school-age children with and without hearing loss. *Journal of Speech, Language and Hearing Research*, 45, 573–584.

Howard, C. S., Munro, K., & Plack, C. J. (2010). Listening effort at signal-to-noise ratios that are typical of the school classroom. *International Journal of Audiology*, 49, 928–932.

Jamieson, D.G., Kranjc, G., & Yu, K. (2004). Speech intelligibility of young school-aged children in the presence of real-life classroom noise. *Journal of the American Academy of Audiology*, 15, 508–517.

McFadden, B., & Pittman, A. (2008). Effect of minimal hearing loss on children's ability to multitask in quiet and in noise. *Language, Speech and Hearing Services in Schools*, 39, 342–351.

Shield, B.M., & Dockrell, J.E. (2003). The effects of noise on children at school: A review. *Journal of Building Acoustics*, 10, 97–106.

Do Hearing Aid Wearers Benefit from Visual Cues?

Wu, Y.H., & Bentler, R.A. (2010). Impact of visual cues on directional benefit and preference: Part I—laboratory tests. *Ear and Hearing, 31*(1), 22–34.

The benefits of directional-microphone use have been consistently supported by experimental data in the laboratory (Gravel et al., 1999; Kuk et al., 1999; Ricketts & Hornsby 2006; Valente et al., 1995). Similarly, hearing aid users have indicated a preference for directional microphones over omnidirectional processing in noise in controlled environments (Amlani et al. 2006; Preves et al., 1999; Walden et al., 2005). Despite the robust directional benefit reported in laboratory studies, field studies have yielded less impressive results, with some studies reporting perceived benefit (Preves et al., 1999; Ricketts et al., 2003) while others have not (Cord et al., 2002, 2004; Palmer ct al., 2006; Walden et al., 2000).

One factor that could account for reduced directional benefit reported in field studies is the availability of visual cues. It is well established that visual cues, including lip-reading (Sumby & Pollack, 1954) as well as eyebrow (Bernstein et al., 1989) and head movements (Munhall et al., 2004), can improve speech-recognition ability in the presence of noise. In field studies, the availability of visual cues could result in a decreased directional benefit due to ceiling effects. In other words, the benefit of audio-visual (AV) speech cues might result in omnidirectional performance so close to a listener's maximum ability that directionality may offer only limited additional improvement. This could reduce both measured and perceived directional benefits. It follows that ceiling effects from the availability of AV speech cues could also reduce the ability of auditory-only (AO) laboratory findings to accurately predict real-world performance.

Few studies have investigated the effect of visual cues on hearing aid performance or directional benefit. Wu and Bentler's goal in the current study was to determine if visual cues could partially account for the discrepancy between laboratory and field studies of directional benefit. They outlined three experimental hypotheses:

1.   Listeners would obtain less directional benefit and would prefer directional- over omnidirectional-microphone modes less frequently in auditory-visual (AV) conditions than in auditory-only (AO) conditions.

2.   The AV directional benefit would not be predicted by the AO directional benefit.

3.   Listeners with greater lip-reading skills would obtain less AV directional benefit than would listeners with lesser lip-reading skills.

Twenty-four adults with hearing loss participated in the study. Participants were between the ages of twenty and seventy-nine years, had bilaterally symmetrical, downward-sloping, sensorineural hearing losses, normal or corrected-normal vision, and were native English speakers. Participants were fitted with bilateral, digital, in-the-ear hearing instruments with manually accessible omnidirectional- and directional-microphone modes.

Directional benefit was assessed with two speech-recognition measures: the AV version of the Connected Speech Test (CST) (Cox et al., 1987) and the Hearing in Noise Test (HINT) (Nilsson et al., 1994). For the AV CST the talker was displayed on a seventeen-inch monitor. Participants listened to sets of CST sentences again in a second session to evaluate subjective preference for directional-versus omnidirectional-microphone modes. Speech stimuli were presented in six signal-to-noise (SNR) conditions ranging from -10dB to +10dB in 4db steps. Lip-reading ability was assessed with the Utley test (Utley, 1946), an inventory of thirty-one sentences recited without sound or facial exaggeration.

Analysis of the CST scores yielded significant main effects for SNR, microphone mode, and presentation mode (AV vs. AO) as well as significant interactions among the variables. The benefit for visual cues was greater than the benefit afforded by directionality. As the authors expected, for most SNRs the directional benefit was smaller for AV conditions than AO conditions with the exception of the poorest SNR condition, -10dB. Scores for all conditions (AV-DIR,

AV-OMNI, AO-DIR, AO-OMNI) plateau at ceiling levels for the most favorable SNRs; meaning that both AV benefit and directional benefit decreased as SNR improved to +10dB. HINT scores, which did not take into account ceiling effects, yielded a significant mean directional benefit of 3.9dB.

Participants preferred the directional-microphone mode in the AO condition, especially at SNRs between -6dB to +2dB. At more favorable SNRs, there was essentially no preference. In the AV condition, participants were less likely to prefer the directional mode, except at the poorest SNR, -10dB. Further analysis revealed that the odds of preferring directional mode in AO condition were 1.37 times higher than in the AV condition. In other words, adding visual cues reduced overall preference for the directional-microphone mode.

At intermediate and favorable SNRs there was no significant correlation between AV directional benefit and the Utley lip-reading scores. For unfavorable SNRs, the negative correlation between these variables was significant, indicating that in the most difficult listening conditions, listeners with better lip-reading skills obtained less AV directional benefit than those participants who were less adept at lip-reading.

The outcomes of these experiments generally support the authors' hypotheses. Visual cues significantly improved speech-recognition scores in omnidirectional trials close to ceiling levels, reducing directional benefit and subjective preference for directional-microphone modes. Auditory-only (AO) performance, typical of laboratory testing, was not predictive of auditory-visual (AV) performance. This is in agreement with prior indications that AO directional benefit as measured in laboratory conditions doesn't match real-world directional benefit and suggests that the availability of visual cues can at least partially explain the discrepancy. The authors suggested that directional benefit should theoretically allow a listener to rely less on visual cues. However, face-to-face conversation is natural and hearing-impaired listeners should leverage avoid visual cues when they are available.

The results of Wu & Bentler's study (2010) suggest that directional microphones may provide only limited additional benefit when visual cues are available, for all but the most difficult listening environments. In the poorest SNRs, directional microphones may be leverages for greater benefit. Still, the authors point out that mean speech-recognition scores were best when both directionality and visual cues were available. It follows that directional microphones should be recommended for use in the presence of competing noise, especially in high-noise conditions. Even if speech-recognition ability is not significantly improved with the use of directional microphones in many typical SNRs, there may be other subjective benefits to directionality, such as reduced listening effort, distraction, or annoyance that listeners respond favorably to.

It is important for audiologists to prepare new users of directional microphones to have realistic expectations. Patients should be advised that directionality can reduce competing noise but not eliminate it. Hearing aid users should be encouraged to consider their positioning relative to competing noise sources and always face the speech source that they wish to attend to. Although visual cues appear to offer greater benefits to speech recognition than directional microphones alone, the availability of visual speech cues may be compromised by poor lighting, glare, crowded conditions, or visual disabilities, making directional microphones all the more important for many everyday situations. Thus all efforts should be made to maximize directionality and the availability of visual cues in day-to-day situations, as both offer potential real-world benefits.

References

Amlani, A.M., Rakerd, B., & Punch, J.L. (2006). Speech-clarity judgments of hearing aid processed speech in noise: differing polar patterns and acoustic environments. *International Journal of Audiology*, 12, 202–214.

Bernstein, L.E., Eberhardt, S.P., & Demorest, M.E. (1989). Single-channel vibrotactile supplements to visual perception of intonation and stress. *Journal of the Acoustical Society of America*, 85, 397–405.

Cord, M.T., Surr, R.K., Walden, B.E. & Olson, L. (2002). Performance of directional microphone hearing aids in everyday life. *Journal of the American Academy of Audiology*, 13, 295–307.

Cord, M.T., Surr, R.K., Walden, B.E.& Dyrlund, O. (2004). Relationship between laboratory measures of directional advantage and everyday success with directional microphone hearing aids. *Journal of the American Academy of Audiology*, 15, 353–364.

Cox, R.M., Alexander, G.C. & Gilmore, C. (1987). Development of the Connected Speech Test (CST). *Ear and Hearing*, 8, 119S–126S.

Gravel, J.S., Fausel, N., Liskow, C., et al. (1999). Children's speech recognition in noise using omnidirectional and dual-microphone hearing aid technology. *Ear and Hearing*, 20, 1–11.

Kuk, F., Kollofski, C., Brown, S., et al. (1999). Use of a digital hearing aid with directional microphones in school-aged children. *Journal of the American Academy of Audiology*, 10, 535–548.

Munhall, K.G., Jones, J.A., Callan, D.E., et al. (2004). Visual prosody and speech intelligibility: head movement improves auditory speech perception. *Psychological Science*, 15, 133–137.

Nilsson, M., Soli, S., & Sullivan, J.A. (1994). Development of the Hearing in Noise Test for the measurement of speech reception thresholds in quiet and in noise. *Journal of the Acoustical Society of America*, 95, 1085–1099.

Palmer, C., Bentler, R., & Mueller, H.G. (2006). Evaluation of a second order directional microphone hearing aid: Part II—Self-report outcomes. *Journal of the American Academy of Audiology*, 17, 190–201.

Preves, D.A., Sammeth, C.A., & Wynne, M.K. (1999). Field trial evaluations of a switched directional/omnidirectional in-the-ear hearing instrument. *Journal of the American Academy of Audiology*, 10, 273–284.

Ricketts, T., & Hornsby, B.W. (2006). Directional hearing aid benefit in listeners with severe hearing loss. *International Journal of Audiology*, 45, 190–197.

Ricketts, T., Henry, P., & Gnewikow, D. (2003). Full time directional versus user selectable microphone modes in hearing aids. *Ear and Hearing*, 24, 424–439.

Sumby, W.H., & Pollack, I. (1954). Visual contribution to speech intelligibility in noise. *Journal of the Acoustical Society of America*, 26, 212–215.

Utley, 1995, missing

Valente, M., Fabry, D.A., & Potts, L.G. (1995). Recognition of speech in noise with hearing aids using dual microphones. *Journal of the American Academy of Audiology, 6*, 440–449.

Walden, B.E., Surr, R.K., Cord, M.T., et al. (2000). Comparison of benefits provided by different hearing aid technologies. *Journal of the American Academy of Audiology, 11*, 540–560.

Walden, B.E., Surr, R.K., Grant, K.W., et al. (2005). Effect of signal-to-noise ratio on directional microphone benefit and preference. *Journal of the American Academy of Audiology, 16*, 662–676.

Wu, Y-H., & Bentler, R.A. (2010). Impact of visual cues on directional benefit and preference: Part I—laboratory tests. *Ear and Hearing, 31*(1), 22–34.

Prescribing Hearing Aids

Recommendations for Fitting the Patient with Cochlear Dead Regions

Cox, R.M., Alexander, G.C., Johnson, J., & Rivera, I. (2011). Cochlear dead regions in typical hearing aid candidates: Prevalence and implications for use of high-frequency speech cues. *Ear & Hearing, 32* (3), 339–348.

Audibility is a well-known predictor of speech-recognition ability (Humes, 2007) and audibility of high-frequency information is of particular importance for consonant identification. Therefore, audibility of high-frequency speech cues is appropriately regarded as an important element of successful hearing aid fittings (Killion & Tillman, 1982; Skinner & Miller, 1983). In contrast to this expectation, some studies have reported that high-frequency gain might have limited or even negative impact on speech-recognition abilities of some individuals (Ching et al., 1998; Hogan & Turner, 1998; Murray & Byrne, 1986). These researchers observed that when high-frequency hearing loss exceeded 55–60dB, some listeners were unable to benefit from increased high-frequency audibility. A potential explanation for this variability was provided by Moore (2001), who suggested that an inability to benefit from amplification in a particular frequency region could be due to cochlear "dead regions," or regions where there is a loss of inner hair cell functioning.

Moore (2001) suggested that hearing aid fittings could potentially be improved if clinicians were able to identify patients with cochlear dead regions (DRs), working under the assumption that diagnosed DRs may contraindicate high-frequency amplification. He and his colleagues developed the TEN test as a method of determining the presence of cochlear dead regions (Moore et al., 2000, 2004). The advent of the TEN test provided a standardized measurement protocol for DRs, but there is still wide variability in the reported prevalence of DRs. Estimates range from as 29% (Preminger et al., 2005) to as high as 84% (Hornsby & Dundas, 2009), with other studies reporting DR prevalence somewhere in the middle of that range. Several potential factors are likely to contribute to this variability, including degree of hearing loss, audiometric configuration, and test technique.

In addition to the variability in reported prevalence of DRs, there is also variability in the reports of how DRs affect the ability to benefit from high-frequency speech cues (Baer et al., 2002; Mackersie et al., 2004; Vickers et al., 2001). It remains unclear as to whether high-frequency amplification recommendations should be modified to reflect the presence of DRs. Most research is in agreement that as hearing thresholds increase, the likelihood of DRs also increases.Patients with severe-to-profound hearing losses are likely to have at least one DR. Because a large proportion of patients have moderate-to-severe hearing losses, Dr. Cox and colleagues (2001) wanted to determine the prevalence of DRs in this population. In addition, they examined the effect of DRs on the use of high-frequency speech cues by individuals with moderate-to-severe loss.

Their study addressed two primary questions:

1)   What is the prevalence of dead regions (DRs) among listeners with hearing thresholds in the 60-90dB range?

2)   For individuals with hearing loss in the 60-90dB range, do those with DRs differ from those without DRs in their ability to use high-frequency speech cues?

One hundred and seventy adults with bilateral, flat, or sloping sensorineural hearing loss were tested. All subjects had thresholds of 60 to 90dB in the better ear for at least part of the range from 1–3kHz and thresholds no better than 25dB for frequencies below 1kHz. Subjects ranged in age from thirty-eight to ninety-six years, and 59% of the subjects had experience with hearing aids.

First, subjects were evaluated for the presence of DRs with the TEN test. Then, speech recognition was measured using high-frequency emphasis (HFE) and low-pass filtered (HFE-LP) stimuli from the QSIN test (Killion et al., 2004). HFE items on this test are amplified up to 32dB above 2.5kHz, whereas the HFE-LP items have much less gain in this range. Comparison of subjects' responses to these two types of stimuli allowed the investigators to assess changes in speech intelligibility with additional high-frequency cues.

Presentation levels for the QSIN were chosen by using a loudness scale and bracketing procedure to arrive at a level that the subject considered "loud but okay." Finally, audibility differences for the two QSIN conditions were estimated using the Speech Intelligibility Index based on ANSI 3.5-1997 (ANSI, 1997).

The TEN test results revealed that 31% of the participants had DRs at one or more test frequencies. Of the 307 ears tested, 23% were found to have a DR for one or more frequencies. Among those who tested positive for DRs, about one-third had DRs in both ears and two-thirds had DRs in one ear or the other in equal proportion. Mean audiometric thresholds were essentially identical for the two groups below 1kHz, but above 1kHz thresholds were significantly poorer for the group with DRs than for the group without DRs. DRs were most prevalent at frequencies above 1.5kHz. There were no age or gender differences.

On the QSIN test, the mean HFE-LP scores were significantly poorer than the mean HFE scores for both groups. There was also a significant difference in performance based on whether or not the participants had DRs. Perhaps more interestingly, there was a significant interaction between the DR group and test stimuli conditions, in that the additional high-frequency information in the HFE stimuli resulted in slightly greater performance gains for the group without DRs than it did for the group with DRs. Furthermore, subjects with one or more isolated DRs were more able to benefit from the high-frequency cues in the HFE lists than were those subjects with multiple, contiguous DRs. Although there were a few isolated individuals who demonstrated lower scores for the HFE stimuli, the differences were not significant and could have been explained by measurement error. Therefore, the authors conclude that the additional high-frequency information in the HFE stimuli was not likely to have had a detrimental effect on performance for these individuals.

As had also been reported in previous studies, subject groups with DRs had poorer mean audiometric thresholds than the groups without DRs, so it was possible that audibility played a role in QSIN performance. Analysis of audibility for QSIN stimuli for the two groups revealed that high-frequency cues in the HFE lists were indeed more audible for the group without DRs. In accounting for this audibility effect, the presence of DRs still had a small but significant effect on performance.

The results of this study suggest that listeners with cochlear DRs still benefit from high-frequency speech cues, albeit slightly less than those without dead regions. The performance improvements were small and the authors caution that it is premature to draw firm conclusions about the clinical implications of this study. Despite the need for further examination, the results of the current study certainly do not support any reduction in prescribed gain for hearing aid candidates with moderate-to-severe hearing losses. The authors acknowledge, however, that because the findings of this and other studies are based on group data, it is possible that specific individuals may be negatively affected by amplification within dead regions. Based on the research to date, this seems more likely to occur in individuals with profound hearing loss who may have multiple, contiguous DRs.

More study is needed to determine the most effective clinical approach to managing cochlear dead regions in hearing aid candidates. Future research should be done with hearing aid users, including for example, the effects of noise on everyday hearing aid performance for individuals with DRs. A study by Mackersie et al. (2004) showed that subjects with DRs suffered more negatives effects of noise than did the subjects without DRs. If there is a convergence of evidence to this effect, then more appropriate recommendations about the use of high-frequency gain, directional microphones, and noise-reduction schemes could be determined as they relate to DRs. For now, the authors of this study recommend that until there are clear criteria to identify individuals for whom high-frequency gain could have deleterious effects, audiologists should continue using best-practice protocols and provide high-frequency gain according to current prescriptive methods.

References

ANSI (1997). *American National Standard Methods for Calculation of the Speech Intelligibility Index* (Vol. ANSI S3.5-1997). New York: American National Standards Institute.

Baer, T., Moore, B. C. J., Kluk, K. (2002). Effects of low pass filtering on the intelligibility of speech in noise for people with and without dead regions at high frequencies. *Journal of the Acoustical Society of America*, 112, 1133–1144.

Ching,T., Dillon, H., & Byrne, D. (1998). Speech recognition of hearing-impaired listeners: Predictions from audibility and the limited role of high-frequency amplification. *Journal of the Acoustical Society of America, 103*, 1128–1140.

Cox, R.M., Alexander, G.C., Johnson, J., & Rivera, I. (2011). Cochlear dead regions in typical hearing aid candidates: Prevalence and implications for use of high-frequency speech cues. *Ear & Hearing, 32*(3), 339–348.

Hogan, C.A., & Turner, C.W. (1998). High-frequency audibility: Benefits for hearing-impaired listeners. *Journal of the Acoustical Society of America, 104*, 432–441.

Hornsby, B. W., & Dundas, J. A. (2009). Factors affecting outcomes on the TEN (SPL) test in adults with hearing loss. *Journal of the American Academy of Audiology*, 20, 251–263.

Humes, L.E. (2007). The contributions of audibility and cognitive factors to the benefit provided by amplified speech to older adults. *Journal of the American Academy of Audiology, 18*, 590–603.

Killion, M. C., Niquette, P. A. & Gudmundsen, G. I. (2004). Development of a quick speech-in-noise test for measuring signal-to-noise ratio loss in normal-hearing and hearing-impaired listeners. *Journal of the Acoustical Society of America*, 116, 2395–2405.

Mackersie, C. L., Crocker, T. L. & Davis, R. A. (2004). Limiting high frequency hearing aid gain in listeners with and without suspected cochlear dead regions. *Journal of the American Academy of Audiology*, 15, 498–507.

Killion, M. C., & Tillman, T.W. (1982). Evaluation of high-fidelity hearing aids. *Journal of Speech and Hearing Research, 25*, 15–25.

Moore, B.C.J. (2001). Dead regions in the cochlea: Diagnosis, perceptual consequences and implications for the fitting of hearing aids. *Trends in Amplification, 5*, 1–34.

Moore, B.C.J., Huss, M., Vickers, D.A., Glasberg, B.R. & Alcantara, J.I. (2000). A test for the diagnosis of dead regions in the cochlea. *British Journal of Audiology, 34*, 205–224.

Preminger, J. E., Carpenter, R. & Ziegler, C. H. (2005). A clinical perspective on cochlear dead regions: Intelligibility of speech and subjective hearing aid benefit. *Journal of the American Academy of Audiology,* 16, 600–613.

Moore, B.C.J., Glasberg, B.R., & Stone, M.A. (2004). New version of the TEN test with calibrations in dB HL. *Ear and Hearing, 25*, 478–487.

Murray, N., & Byrne, D. (1986). Performance of hearing-impaired and normal hearing listeners with various high-frequency cut-offs in hearing aids. *Australian Journal of Audiology, 8*, 21–28.

Skinner, M.W., & Miller, J.D. (1983). Amplification bandwidth and intelligibility of speech in quiet and noise for listeners with sensorineural hearing loss. *Audiology, 22*, 253–279.

Cochlear Dead Regions and High-frequency Gain: How to Fit the Hearing Aid

Cox, R.M., Johnson, J.A. & Alexander, G.C. (2012). Implications of high-frequency cochlear dead regions for fitting hearing aids to adults with mild to moderately severe hearing loss. *Ear and Hearing*, May 2012, e-published ahead of print.

Cochlear dead regions (DRs) are defined as a total loss of inner hair cell function across a limited region of the basilar membrane (Moore, et al., 1999b). This does not result in an inability to perceive sound at a given frequency range, rather the sound is perceived via a spread of excitation to adjacent regions in the cochlea where the inner hair cells are still functioning. Because the response is spread out over a broader tonotopic region, patients with cochlear dead regions may perceive some high-frequency pure tones as "clicks," "buzzes," or "whooshes" rather than tones.

Dead regions can be present at moderate hearing thresholds (e.g. 60dBHL) and are more likely to be present as the degree of loss increases. Psychophysical tuning curves are the preferred method for identifying cochlear dead regions in the laboratory. Moore and his colleagues developed the Threshold Equalizing Noise (TEN) Test as a clinical means of identifying dead regions (Moore et al., 2000; Moore et al., 2004). The TEN procedure looks for shifts in masked thresholds beyond what would typically be expected for a given hearing loss. A threshold obtained with TEN masking noise that shifts at least 10dB indicates the likely presence of a cochlear dead region.

Historically, there has been a lack of consensus among clinical investigators as to whether or not high-frequency gain is beneficial for hearing aid users who have cochlear dead regions. Some studies suggest that high-frequency gain could have deleterious effects on speech perception and should be limited for individuals with cochlear dead regions (Moore, 2001b; Turner, 1999; Padilha et al., 2007). For example, Vickers et al. (2001) and Baer et al. (2002) studied the benefit of high-frequency amplification in quiet and noise for individuals with and without DRs. Both studies reported that individuals with DRs were unable to benefit from high-frequency amplification. While Gordo and Iorio (2007) found that hearing aid users with DRs performed worse with high-frequency amplification than they did without it.

In contrast, Cox and her colleagues (2011) found beneficial effects of high-frequency audibility whether or not the participants had dead regions. Others have reported equivalent performance for participants with and without dead regions for quiet and low-noise conditions; however, in high-noise conditions the individuals without dead regions demonstrated further improvement when additional high-frequency amplification was provided, whereas participants with dead regions did not (Mackersie et al., 2004).

The current study was undertaken to examine the subjective and objective effect of high-frequency amplification on matched pairs of participants (with and without DRs) in a variety of conditions. Participants were fitted with hearing aids that had two programs: the first (NAL) was based on the NAL-NL1 formula and the second (LP) was identical to the NAL-NL1 program below 1000Hz, with amplification rolled off above 1000Hz. The goals of the study were to compare performance with these two programs, for individuals with and without dead regions. The following measures were conducted:

1)   Speech discrimination in quiet laboratory conditions

2)   Speech discrimination in noisy laboratory conditions

3)   Subjective performance in everyday situations

4)   Subjective preference for everyday situations

Participants were recruited from a pool of individuals who had previously been identified as typical hearing aid patients (Cox et al., 2011). Participants had bilateral flat or sloping sensorineural hearing loss with thresholds above 25dB below 1kHz and thresholds of 60 to 90dB HL for at least part of the frequency range of 1–3kHz.

The TEN test (Moore et al., 2004) was administered to determine the presence of DRs. To be eligible for the study, participants needed to have one or more DRs in the better ear at or above 1kHz and no

DRs below 1kHz. Participants were then divided into to two groups: the experimental group with DRs and the control group without DRs. Individuals in the experimental group showed a diverse range of DR distribution across frequency. Almost half of the participants had DRs between 1–2kHz, whereas the remainder had DRs only at or above 3kHz. A little more than half of the participants had one DR only, whereas the others had more than one DR.

Individuals in the experimental group were matched in pairs with individuals from the control group. In total, there were eighteen participant pairs—each matched for age, degree of hearing loss and slope of hearing loss. There were twenty-four men and twelve women. No attempt was made to match pairs based on gender.

Participants were fitted monaurally with behind-the-ear hearing aids coupled to vented skeleton earmolds. The monaural fitting was chosen to avoid complications when participants switched between the NAL and LP programs. Data collection was completed before the widespread availability of wireless hearing aids, so the participants would have had to reliably switch both hearing aids individually to the proper program every time to avoid making occasional subjective judgments based on mismatched programs.

The hearing aids had two programs: a program based on the NAL-NL1 prescription (NAL) and a program with high-frequency roll off (LP). Participants were able to switch the programs themselves but could not identify the programs as NAL or LP. Half of the participants had NAL in P1 and LP in P2, whereas the other half had LP in P1 and NAL in P2. Verification measures were conducted to ensure that the two programs matched below 1kHz and to make sure the participants judged the programs to be equally loud.

After a two-week acclimatization period, participants returned for speech-recognition testing and field-trial training. Speech and noise stimuli were presented in a sound field with the unaided ear plugged during testing. Speech recognition in quiet was evaluated using the Computer Assisted Speech Perception Assessment (CASPA; Mackersie et al., 2001). The CASPA test includes lists of ten consonant-vowel-consonant words spoken by a female. Five lists were presented for each of the NAL and LP programs. Stimuli were presented at 65dB SPL.

Speech recognition in noise was evaluated with the Bamford-Kowell-Bench Speech in Noise (BKB-SIN test, Etymotic Research, 2005), which contains sentences spoken by a male talker, masked by four-talker babble. The test contains lists of ten sentences with thirty-one scoring words. In each list, the signal-to-noise ratio (SNR) decreases by 3dB with each sentence, so that within any list the SNR ranges from +21dB to -6dB. Sentences were presented at 73dB, a "loud but OK" level, as recommended for this test.

Following the speech-recognition testing, participants were trained in the field-trial procedures for subjective ratings. They were asked to evaluate their ability to understand speech in everyday situations with the NAL and LP programs and identify occasions during which they felt they could understand some but not all of the words they were hearing. Participants were given booklets with daily rating sheets and listening checklists to record daily hours of hearing aid use and track the variety of their daily listening experiences.

After a two-week field trial, participants returned to the laboratory for a second session of CASPA and BKB-SIN testing. They submitted ratings sheets and listening checklists and were interviewed about their preferred hearing aid program for everyday listening. The interview consisted of questions that covered program preferences related to: understanding speech in quiet, understanding speech in noise, hearing over long distances, the sound of their own voice, sound quality, loudness, localization, the least tiring program, and the one that provided the most comfortable sound. Participants were asked to indicate their preferred program for each of these criteria, as well as their preferred program for overall everyday use. They were asked to provide three reasons for overall preference.

Speech-recognition testing in quiet revealed no difference in overall performance between the two groups, but there was a significant difference based on the hearing aid program that was used. Listeners from both the experimental group and the control group performed better with the broadband NAL program, though the difference between the NAL and LP programs was larger for the control group than the experimental group. This indicates that the individuals without DRs were able to derive more benefit from the additional high-frequency information in the NAL program than the individuals with DRs did.

Speech-recognition testing in noise revealed a similar finding but in this case the improvement with the NAL program was only significant for the control group. Although the experimental group's mean scores with the NAL program were higher than those with the LP program, the difference did not reach statistical significance.

Because the BKB-SIN test used variable SNR levels, performance-intensity functions were constructed with scores obtained using the NAL and LP programs. These functions revealed that at any given SNR, speech was more intelligible with the NAL program. However, there was more of a difference between the NAL and LP functions for the control group than the experimental group, consistent with a program by group statistical interaction.

Subjective ratings of speech understanding revealed no significant difference between the experimental and control groups, but there was a significant difference based on program. Participants from the control and experimental groups rated their performance better with the NAL program.

Interviews concerning program preference revealed that twenty-three participants preferred the NAL program and eleven preferred the LP program. There was no association with the presence of DRs. When the reasons supporting the participants' preferences were analyzed, the most frequently mentioned reason for NAL preference was greater speech clarity. The most common reason for LP preference was that the other program (NAL) was too loud.

This investigation by Dr. Cox and her colleagues indicates that high-frequency amplification was beneficial to participants with single or multiple DRs, especially for speech recognition in quiet. In noise, participants with DRs still performed better with the NAL program, though the improvement was not as marked as it was for those without DRs. In field trials, participants with DRs reported more improvement with the NAL program than the control group did, indicating that perceived benefits in everyday situations exceeded any predictions of the laboratory results. **At no point in the study did high-frequency amplification reduce performance for individuals with or without DRs.**

This finding is in contrast with previous reports (Vinay & Moore, 2007a; Gordo & Iorio, 2007). Cox and her colleagues note that most of the participants in their study had only one or two DRs as opposed to several contiguous DRs. They allow that their findings might not relate to the performance of participants with several contiguous DRs, but point out that among typical hearing aid candidates, it is unlikely for individuals to have more than one or two DRs. **With this consideration, the authors suggest that high-frequency amplification should not be reduced, even in cases with identified dead regions.**

This study from the University of Memphis provides a recommendation for use of prescribed settings and against reduction of high-frequency gain for hearing aid users with one or two DRs. They found beneficial effects of high-frequency amplification in laboratory and everyday environments and noted no circumstances in which listeners demonstrated deleterious effects of high-frequency amplification. These results may not pertain to individuals with several contiguous DRs but they are pertinent to the majority of typical hearing aid wearers. Their findings also support the use of subjective performance measures, as these provided additional information that was sometimes in contrast to the laboratory results. They point out that laboratory results do not always predict performance in everyday life and it can be extrapolated that clinical measures of efficacy should always be supported with subjective reports of effectiveness, like self-assessment of comfort and acceptance.

References

Baer, T., Moore, B.C. & Kluk, K. (2002). Effects of low pass filtering on the intelligibility of speecdh in noise for people with and without dead regions at high frequencies. *Journal of the Acoustical Society of America,* 112(3 pt. 1), 1133–1144.

Ching, T.Y., Dillon, H. & Byrne, D. (1998). Speech recognition of hearing-impaired listeners: predictions from audibility and the limited role of high-frequency amplification. *Journal of the Acoustical Society of America,* 103, 1128–1140.

Cox, R. M., Alexander, G.C., & Johnson, J.A. (2011). Cochlear dead regions in typical hearing aid candidates: prevalence and implications for use of high-frequency speech cues. *Ear and Hearing,* 32, 339–348.

Cox, R.M., Johnson, J.A. & Alexander, G.C. (2012). Implications of high-frequency cochlear dead regions for fitting hearing aids to adults with mild to moderately severe hearing loss. *Ear and Hearing,* May 2012, e-published ahead of print.

Etymotic Research (2005). *BKB-SIN Speech in Noise Test, Version 1.03.* Elk Grove Village, IL: Etymotic Research.

Moore, B.C., Glasberg, B. & Vickers, D.A. (1999b). Further evaluation of a model of loudness perception applied to cochlear hearing loss. *Journal of the Acoustical Society of America,* 106, 898–907.

Moore, B.C., Huss, M. & Vickers, D.A. (2000). A test for the diagnosis of dead regions in the cochlea. *British Journal of Audiology,* 34, 205–224.

Moore, B.C., Glasberg, B. R. & Stone, M.A. (2004). New version of the TEN test with calibrations in dB HL. *Ear and Hearing,* 25, 478–487.

Gordo, A. & Iorio, M.C. (2007). Dead regions in the cochlea at high frequencies: Implications for the adaptation to hearing aids. *Revista Brasileira de Otorrinolaringologia,* 73, 299–307.

Hogan, C.A. & Turner, C.W. (1998). High frequency audibility: benefits for hearing-impaired listeners. *Journal of the Acoustical Society of America,* 104, 432–441.

Mackersie, C.L., Boothroyd, A. & Minniear, D. (2001). Evaluation of the Computer-Assisted-Speech Perception Assessment Test (CASPA). *Journal of the American Academy of Audiology,* 12, 390–396.

Mackersie, C.L., Crocker, T.L. & Davis, R.A. (2004). Limiting high-frequency hearing aid gain in listeners with and without suspected cochlear dead regions. *Journal of the American Academy of Audiology,* 15, 498–507.

Moore, B.C. (2001a). Dead regions in the cochlear: Diagnosis, perceptual consequences and implications for the fitting of hearing aids. *Trends in Amplification,* 5, 1–34.

Moore, B.C., (2001b). *Dead regions in the cochlear: Implications for the choice of high-frequency amplification.* In R.C. Seewald & J.S. Gravel (Eds). *A Sound Foundation Through Early Amplification,* 153–166. Stafa, Switzerland: Phonak AG.

Padilha, C., Garcia, M.V., & Costa, M.J. (2007). Diagnosing cochlear dead regions and its importance in the auditory rehabilitation process. *Brazilian Journal of Otolaryngology,* 73, 556–561.

Turner, C.W. (1999). The limits of high-frequency amplification. *Hearing Journal,* 52, 10–14.

Turner, C.W. & Cummings, K.J. (1999). Speech audibility for listeners with high-frequency hearing loss. *American Journal of Audiology*, 8, 47–56.

Vickers, D.A., Moore, B.C. & Baer, T. (2001). Effects of low-pass filtering on the intelligibility of speech in quiet for people with and without dead regions at high frequencies. *Journal of the Acoustical Society of America*, 110, 1164–1175.

Vinay, B.T. & Moore, B.C. (2007a). Prevalence of dead regions in subjects with sensorineural hearing loss. *Ear and Hearing*, 28, 231–241.

Understanding the Benefits of Bilateral Hearing Aid Use

Boymans, M., Goverts, S.T., Kramer, S.E., Festen, J.M., & Dreschler, W.A. (2008). A prospective multi-centre study of the benefits of bilateral hearing aids. *Ear and Hearing, 29*(6), 930–941.

The benefits of binaural amplification are generally well established and include improved speech discrimination in noise (Hawkins & Yacullo, 1984; Kobler & Rosenhall, 2002), improved localization of sound sources (Dreschler & Boymans, 1994; Punch et al., 1991), perception of balanced hearing, improved speech clarity (Chung & Stephens, 1986; Erdman & Sedge, 1981), and reduced listening effort (Noble, 2006). However, some studies have shown either little subjective difference between unilateral and bilateral amplification (Andersson et al., 1996) or even a subjective preference for unilateral hearing aids, especially in noise (Schreurs & Olsen, 1985; Walden & Walden, 2005).

The authors of the current study sought to confirm subjective evaluations of binaural hearing aids with objective, functional tests of localization and speech discrimination in noise. They also examined three diagnostic measures to determine their potential as predictors of binaural success.

Two hundred fourteen hearing-impaired subjects were recruited from eight audiology clinics in the Netherlands. Participant inclusion criteria were limited only to participants who were native Dutch speakers and were physically able to complete the test procedures, with no contraindications for binaural hearing aid fitting. Therefore, individual characteristics varied widely with regard to prior hearing aid use, hearing aid style and circuitry, age, and degree of hearing loss. Ten participants with normal hearing were also tested for reference purposes.

Prior to hearing aid fitting, in addition to basic diagnostic audiometry, participants completed three tests that were chosen as potential predictors of binaural benefit:

1.   Interaural time differences
2.   Binaural masking-level differences
3.   Speech-reception thresholds in background noise

Following the hearing aid fittings, functional binaural benefit was evaluated and questionnaires were administered to obtain subjective responses to unilateral and bilateral fittings. Three assessment tools were used:

1.   Speech intelligibility in background noise with spatial separation of speech and noise
2.   Horizontal localization of everyday sounds
3.   Subjective questionnaires to examine differences between unaided, unilateral, and bilateral conditions for detection of sounds, discrimination of sounds, speech intelligibility in quiet and noise, localization, and comfort of loud sounds

Not surprisingly, on all three diagnostic measures, normal hearing participants performed significantly better than hearing-impaired participants. There was a great deal of inter-participant variability within the hearing-impaired group.

On the functional test of speech intelligibility with spatially separated speech and noise, bilateral hearing aid users performed significantly better than unilateral hearing aid users. Improvements were noted for conditions in which competing sounds were presented ipsilateral and contralateral to the speech stimulus. On the localization test, bilateral hearing instrument wearers again performed significantly better than unilateral hearing aid wearers. Subjective questionnaires showed that unilateral hearing aid use was favored over unaided conditions for all categories except comfort of loud sounds. Similarly, bilateral hearing aid use was favored over unilateral for all categories except comfort of loud sounds. This finding is in agreement with previous work by the lead author of the current study (Boymans, 2003).

Participants were asked to provide reasons why they preferred one or two hearing aids. The most common reason for preferring a unilateral fitting was that the user's own voice was more pleasant with one hearing aid. For preferred bilateral fittings, the most common reasons were, intelligibility on both sides, better localization, better sound quality, and better balance. Following completion of the study, 93% of the participants chose to purchase bilateral hearing aids, whereas 7% chose to purchase only one hearing aid.

One primary goal of the study was to determine if subjective benefit could be supported with objective test results. There was a significant positive correlation between bilateral benefit for speech understanding and subjective satisfaction ratings, but other evaluated factors did not show this relationship. Therefore, the authors determined that functional test results could not distinguish between groups who preferred unilateral or bilateral fittings. Overall, however, the vast majority of participants preferred bilateral hearing aid fittings and the functional test results support a strong binaural benefit.

The second goal of the study was to evaluate potential predictive measures of binaural benefit. The results did not show strong correlations between bilateral hearing aid performance and interaural time difference, binaural masking level difference or speech reception threshold measures. Therefore, these measures were determined not to have particular predictive value for determining binaural hearing aid success. In fact, the strongest correlation between bilateral benefit and any other diagnostic measure was found for traditional audiometric measures of pure-tone average and speech-recognition ability.

Binaural benefit was also examined with regard to other subject variables. The authors found greater binaural benefit for users with more severe hearing loss and for those with more symmetrical hearing loss. There were no significant differences between subjects who had previously been fitted with unilateral hearing aids and those who had been previously fitted bilaterally. Participants without prior hearing aid experience demonstrated slightly less binaural benefit and less satisfaction than those with previous experience. The authors point out that this finding is confounded by the fact that previous users tended to have significantly greater degrees of hearing loss than first-time users.

The bilateral benefit for localization was higher for in-the-ear hearing aid users than for behind-the-ear hearing aid users. The authors surmised that this could be related to pinna effects, but pinna effects generally aid vertical localization and front/back localization (Blauert, 1997), whereas the localization measures in the current study were strictly horizontal. Still, it is possible that preservation of pinna-related spectral cues in combination with binaural cues could have had an additive effect for the in-the-ear hearing aid users in the present study.

It is interesting to note that despite the highly variable subject population in this study, significant binaural benefit for speech intelligibility and localization was found across participants, and participants overwhelmingly preferred the use of binaural hearing aids over monaural. Variables such as microphone mode, noise-reduction technology, and circuit quality were not specifically addressed or controlled. It is reasonable to surmise that performance in the one category in which subjects preferred unilateral hearing aids, comfort for loud sounds, could be improved by adjustments to noise-reduction settings, MPO or gain settings, or use of adaptive directionality. Therefore, the study as a whole offers strong support for binaural hearing aid recommendations and indicates that the only negative effect, that of loudness discomfort, could probably be easily corrected with current technology.

The binaural benefits measured in this study can probably be reasonably extrapolated to individuals with asymmetrical hearing losses, but this issue might benefit from further study. Also, it is likely that similar binaural benefits may also apply to potential hearing aid users who are unwilling or reluctant to consider binaural hearing aid use, but these patients will require more thorough counseling with regard to expectations and acclimatization. The primary reason given for unilateral hearing aid preference was related to occlusion and the sound quality of one's own voice. A reluctant user of new binaural hearing aids will need to understand that this is a common, but often short-lived outcome of binaural hearing aid use.

Because of the poor predictive value of diagnostic tests for binaural hearing aid success, the authors advise that it is probably best for patients to determine binaural benefit for themselves during their initial trial period. This is appropriate advice and may be in line with what most clinicians are already recommending to their patients. Because an individual's work, home, and social activities are important determinants of his/her perceived hearing handicap, binaural hearing aids should always be tested thoroughly in these situations to evaluate benefit. There is little financial risk involved, as most clinics offer at least a thirty-day trial period with new instruments and many offer a forty-five- or sixty-day trial. Should a client determine that the benefit of a second hearing aid does not outweigh the financial burden, they would be able to return the aid for a refund, losing only the cost of a custom earmold and/or a trial-period fee.

The current study shows strong evidence for functional improvements as well as perceived advantages in binaural hearing aid users. However, the authors were unable to identify a diagnostic tool to effectively predict binaural success. This raises an important question about the value of such a predictive measure. The significant improvements enjoyed by binaural users and the overwhelming preference for two hearing aids over one suggest that binaural fittings should be the recommendation of choice for all clients with bilateral, aidable hearing loss. Granted, there are some audiometric findings that preclude a binaural recommendation, such as profound hearing loss in one ear, normal hearing in one ear, or exceptionally poor word-recognition ability in one ear. But these are obvious, well-known, and relatively uncommon clinical contraindications to binaural hearing aid use. It seems reasonable, as the authors eventually suggest, to forgo predictive measures and allow clients to experience binaural benefits individually and determine the proper decision for themselves during their trial period.

References

Andersson, G., Palmkvist, A., & Melin, L. (1996). Predictors of daily assessed hearing aid use and hearing capability using visual analogue scales. *British Journal of Audiology*, 30, 27–35.

Blauert, J. (1997). *Spatial Hearing: The Psychophysics of Human Sound Localization.* Cambridge: MIT Press.

Boymans, M. (2003). *Intelligent processing to optimize the benefits of hearing aids.* (Unpublished doctoral dissertation). University of Amsterdam.

Boymans, M., Goverts, S.T., Kramer, S.E., Festen, J.M., & Dreschler, W.A. (2008). A prospective multi-centre study of the benefits of bilateral hearing aids. *Ear and Hearing*, 29(6), 930–941.

Chung, S.M., & Stephens, S.D. (1986). Factors influencing binaural hearing aid use. *British Journal of Audiology*, 20, 129–140.

Dreschler, W.A., & Boymans, M. (1994). Clinical evaluation of the advantage of binaural hearing aid fittings. *Audiologische Akustik*, 5, 12–23.

Erdman, S.A., & Sedge, R.K. (1981). Subjective comparisons of binaural versus monaural amplification. *Ear and Hearing*, 2, 225–229.

Hawkins, D.B., & Yacullo, W.S. (1984). Signal-to-noise ratio advantage of binaural hearing aids and directional microphones under different levels of reverberation. *Journal of Speech and Hearing Disorders*, 49, 278–186.

Kobler, S., & Rosenhall, U. (2002). Horizontal localization and speech intelligibility with bilateral and unilateral hearing aid amplification. *International Journal of Audiology*, 41, 395–400.

Noble, W. (2006). Bilateral hearing aids: a review of self-reports of benefit in comparison with unilateral fitting. *International Journal of Audiology*, 45, 63–71.

Punch, J.L., Jenison, R.L., & Alan, J. (1991). Evaluation of three strategies for fitting hearing aids binaurally. *Ear and Hearing*, 12, 205–215.

Schreurs, K.K., & Olsen, W.O. (1985). Comparison of monaural and binaural hearing aid use on a trial period basis. *Ear and Hearing*, 6, 198–202.

Walden, T.C.. & Walden, B.E. (2005). Unilateral versus bilateral amplification for adults with impaired hearing. *Journal of the American Academy of Audiology*, 16, 574–584.

True or False? Two Hearing Aids Are Better Than One

McArdle, R., Killion, M., Mennite, M. & Chisolm, T. (2012). Are Two Ears Not Better Than One? *Journal of the American Academy of Audiology*, *23*, 171–181.

Audiologists are accustomed to recommending two hearing aids for individuals with bilateral hearing loss, based on the known benefits of binaural listening (Carhart, 1946; Keys, 1947; Hirsh, 1948; Koenig, 1950), though the potential advantages of binaural versus monaural amplification have been debated for many years.

One benefit of binaural listening, binaural squelch, occurs when the signal and competing noise come from different directions (Kock, 1950; Carhart, 1965). When the noise and signal come from different directions, time and intensity differences cause the waveforms arriving at each ear to be different, resulting in a dichotic listening situation. The central auditory system is thought to combine these two disparate waveforms and essentially subtract the waveform arriving at one side from that of the other, resulting in an effective SNR improvement of about 2–3dB (Dillon, 2001).

Binaural redundancy, another potential benefit of listening with two ears, is an advantage created simply by receiving similar information in both ears. Dillon (2001) describes binaural redundancy as allowing the brain to get two "looks" at the same sound, resulting in SNR improvement of another 1–2 dB (MacKeith & Coles, 1971; Bronkhorst & Plomp, 1988).

Though the benefits of binaural listening would imply benefits of binaural amplification as well, there has been a lack of consensus among researchers. Some studies have reported clear advantages to binaural amplification over monaural fittings, but others have not. Decades ago a number of studies were published on both sides of the argument, but differences in outcomes may have been related to speaker location and the presentation angles of the speech and noise signals (Ross, 1980) so the potential advantages of binaural amplification were still unclear.

Some recent reports have supported the use of monaural amplification over binaural for some individuals, in objective and subjective studies. Henkin et al. (2007) reported that 71% of their subjects performed better on a speech-in-noise task when fitted with one hearing aid on the "better" ear than when fitted with two hearing aids. Cox et al. (2011) reported that 46% of their subjects preferred to use one hearing aid rather than two.

In contrast, a report by Mencher & Davis (2006) concluded that 90% of adults perform better with two hearing aids. They explained that 10% of adults may have experienced negative binaural interaction or binaural interference, which is described as the inappropriate fusion of signals received at the two ears (Jerger et al., 1993; Chmiel et al., 1997).

The phenomenon of binaural interference and the potential advantage of monaural amplification was investigated by Walden and Walden (2005). In a speech-recognition-in-noise task in which speech and the competing babble were presented through a single loudspeaker at 0-degrees azimuth, they found that performance with one hearing aid was better than binaural for 82% of their participants. This is in contrast to Jerger's (1993) report of an incidence of 8–10% for subjects that might have experienced binaural interference. One criticism of Walden and Walden's study is that their "monaural" condition left the unaided ear open. Their presentation level of 70dB HL and the use of subjects with mild-to-moderate hearing loss indicates that subjects were still receiving speech and noise cues in the unaided ear, resulting in an albeit modified, binaural listening situation. Furthermore, their choice of one single loudspeaker for presentation of noise and speech directly in front of the listener created a diotic listening condition, which eliminated the use of binaural head shadow cues. This methodology may have limited their study's relevance to typical everyday situations in which listeners are engaged in face-to-face conversation with competing noise all around.

Because the potential advantages or disadvantages of binaural amplification have such important clinical implications, Rachel McArdle and her colleagues sought to clarify the issue with a two-part study of monaural and binaural listening. The first experiment was an effort to replicate Walden and Walden's 2005 sound-field study, this time adding a true monaural condition and an unaided condition. The second experiment examined monaural versus diotic and dichotic listening conditions, using real-world recordings from a busy restaurant.

Twenty male subjects were recruited from the Bay Pines Veteran's Affairs Medical Facility. Subjects ranged in age from fifty-nine to eighty-five years old and had bilateral, symmetrical hearing losses. All were experienced users of binaural hearing aids.

For the first experiment, subjects wore their own hearing aids, so a variety of models from different manufacturers were represented. Hearing aids were fitted according to NAL-NL1 prescriptive targets and were verified with real-ear measurements. All of the hearing aids were multi-channel instruments with directional microphones, noise reduction, and feedback management. None of the special features were disabled during the study.

Subjects were tested in sound field, with a single loudspeaker positioned three feet in front of them. They were tested under five conditions: 1) right ear aided, left ear open, 2) left ear aided, right ear open, 3) binaurally aided, 4) right ear aided, left ear plugged (true monaural), and 5) unaided. The QuickSIN test (Killion et al., 2004) was used to evaluate sentence recognition in noise in all of these conditions. The QuickSIN test yields a value for "SNR loss," which represents the SNR required to obtain a score of 50% correct for key words in the sentences.

The primary question of interest for the first experiment asked whether two aided ears would achieve better performance than one aided ear. The results showed that only 20% of the participants performed better with one aid, whereas 80% performed better with binaural aids. The lowest SNR loss values were for the binaural condition, followed by right ear aided, left ear aided, true monaural (with left ear plugged), and unaided. Analysis of variance revealed that the binaural condition was significantly better than all other conditions. The right-ear only condition was significantly better than unaided, but all other comparisons failed to reach significance.

The results of Experiment 1 are comparable to results reported by Jerger (1993) but contrast sharply with Walden and Walden's 2005 study, in which 82% of respondents performed better monaurally aided. To compare their results further to Walden and Walden's, McArdle and her colleagues compiled scores for the subjects' better ears and found that there was no significant difference between binaural and better-ear performance, but both of these conditions were significantly better than the other conditions. They also examined the effect of degree of hearing loss and found that individuals with hearing thresholds poorer than 70dB HL were able to achieve about twice as much improvement from binaural amplification as those subjects with better hearing. Still, the results of Experiment 1 support the benefit of binaural hearing aids for most participants, especially those with poorer hearing.

The purpose of Experiment 2 was to further examine the potential benefit of hearing with two ears, using diotic and dichotic listening conditions. Diotic listening refers to a condition in which the listener receives the same stimulus in both ears, whereas dichotic listening refers to more typical real-world conditions in which each ear receives slightly different information, subject to head shadow and time and intensity differences.

Speech recognition was evaluated in four conditions: 1) monaural right, 2) monaural left, 3) diotic, and 4) binaural or dichotic. Materials for the R-SPACE QSIN test (Revit, et al., 2007) were recorded through a KEMAR manikin with competing restaurant noise presented through eight loudspeakers. Recordings were taken from eardrum-position microphones on each side of KEMAR, thus preserving binaural cues that would be typical for a listener in a real-world setting.

In Experiment 2, subjects were not tested wearing hearing aids; the stimuli were presented via inserted earphones. The use of recorded stimuli presented under earphones eliminated the potentially confounding factor of hearing aid technology on performance and allowed the presentation of real-world recordings in truly monaural, diotic, and dichotic conditions.

The best performance was demonstrated in the binaural condition, followed by the diotic condition, then the monaural conditions. The binaural condition was significantly better than diotic, and both the diotic and dichotic conditions were significantly better than the monaural conditions. Again in contrast to Walden and Walden's study, 80% of the subjects scored better in the binaural condition than either of the monaural conditions and 65% of the subjects scored better in the diotic condition than either monaural condition. These results support the findings of the first experiment and indicate that for the majority of listeners, speech recognition in noise improves when two ears are listening instead of one.

Furthermore, the finding that the binaural condition was significantly better than the diotic condition indicates that it is not only the use of two ears, but also the availability of binaural cues that have a positive impact on speech recognition in competing noise.

McArdle and her colleagues point out that their study, as well as other recent reports (Walden & Walden, 2005; Henkin, 2007), suggests that the majority of listeners perform better on speech-in-noise tasks when they are listening with two ears. When binaural time and intensity cues are available, performance is even better than with the same stimulus reaching each ear. They also found that the potential benefit of binaural hearing was even more pronounced for individuals with more severe hearing loss. This supports the recommendation of binaural hearing aids for individuals with bilateral hearing loss, especially those with severe loss.

Cox et al. (2011) reported that listeners who experienced better performance in everyday situations tended to prefer binaural hearing aid use, but also found that forty-three out of ninety-four participants generally preferred monaural to binaural use over a twelve-week trial. For new hearing aid users or prior monaural users, this is not surprising, as it can take time to adjust to binaural hearing aid use. Clinical observation suggests that individuals who have prior monaural hearing aid experience may have more difficulty adjusting to binaural use than individuals who are new to hearing aids altogether. However, with consistent daily use, reasonable expectations and appropriate counseling, most users can successfully adapt to binaural use. It is possible that the subjects in Cox et al.'s study who preferred monaural use were responding to factors other than performance in noise. If they were switching between monaural and binaural use, perhaps they did not wear the two instruments together consistently enough to fully acclimate to binaural use in the time allotted.

Though their study presented strong support for binaural hearing aid use, McArdle and her colleagues suggest that listeners may benefit from "self-experimentation" to determine the optimal configuration with their hearing aids. This suggestion is an excellent one, but it may be most helpful within the context of binaural use. Even patients with adaptive and automatic programs can be fitted with manually accessible programs designed for particularly challenging situations and should be encouraged to experiment with these programs to determine the optimal settings for their various listening needs.

Clinicians who have been practicing for several years may recall the days when hearing aid users often lost their hearing aids in restaurants because they had removed one aid in order to more easily ignore background noise. That is less likely to occur now, as current technology can help most hearing aid users function quite well in noisy situations. With directional microphones and multiple programs, along with the likelihood that speech and background noise are often spatially separated, binaural hearing aids are likely to offer advantageous performance for speech recognition in most acoustic environments. Bilateral data exchange and wireless communication offer additional binaural benefits, as two hearing instruments can work together to improve performance in noise and provide binaural listening for telephone or television use.

References

Bronkhorst, A.W. & Plomp, R. (1988). The effect of head induced interaural time and level differences on speech intelligibility in noise. *Journal of the Acoustical Society of America*, 83, 1508–1516.

Carhart, R. (1965). Problems in the measurement of speech discrimination. *Archives of Otolaryngology*, 82, 253–260.

Carhart, R. (1946). Selection of hearing aids. *Archives of Otolaryngology*, 44, 1–18.

Chmiel, R., Jerger, J., Murphy, E., Pirozzolo, R. & Tooley, Y.C. (1997). Unsuccessful use of binaural amplification by an elderly person. *Journal of the American Academy of Audiology*, 8, 1–10.

Cox, R.M., Schwartz, K.S., Noe, C.M. & Alexander, G.C. (2011). Preference for one or two hearing aids among adult patients. *Ear and Hearing*, 32 (2), 181–197.

Dillon, H. (2001). Monaural and binaural considerations in hearing aid fitting. In: Dillon, H., ed.*Hearing Aids*. Turramurra, Australia: Boomerang Press, 370–403.

Henkin, Y., Waldman, A. & Kishon-Rabin, L. (2007). The benefits of bilateral versus unilateral amplification for the elderly: are two always better than one? *Journal of Basic and Clinical Physiology and Pharmacology*, 18(3), 201–216.

Hirsh, I.J. (1948). Binaural summation and interaural inhibition as a function of the level of masking noise. *American Journal of Psychology*, 61, 205–213.

Jerger, J., Silman, S., Lew, J. & Chmiel, R. (1993). Case studies in binaural interference: converging evidence from behavioral and electrophysiologic measures. *Journal of the American Academy of Audiology*, 4, 122–131.

Keys, J.W. (1947). Binaural versus monaural hearing. *Journal of the Acoustical Society of America*, 19, 629–631.

Killion, M.C., Niquette, P.A., Gudmundsen, G.I., Revit, L.J. & Banerjee, S. (2004). Development of a quick speech-in-noise test for measuring signal-to-noise ratio loss in normal hearing and hearing-impaired listeners. *Journal of the Acoustical Society of America*, 116, 2395–2405.

Kock, W.E. (1950). Binaural localization and masking. *Journal of the Acoustical Society of America*, 22, 801–804.

Koenig, W. (1950). Subjective effects in binaural hearing. [Letter to the Editor] *Journal of the Acoustical Society of America*, 22, 61–62.

MacKeith, N.W. & Coles, R.A. (1971). Binaural advantages in hearing speech. *Journal of Laryngology and Otology*, 85, 213–232.

McArdle, R., Killion, M., Mennite, M. & Chisolm, T. (2012). Are Two Ears Not Better Than One? *Journal of the American Academy of Audiology*, 23, 171–181.

Mencher, G.T. & Davis, A. (2006). Binaural or monaural amplification: is there a difference? A brief tutorial. *International Journal of Audiology*, 45, S3–S11.

Revit, L., Killion, M. & Compton-Conley, C. (2007). Developing and testing a laboratory sound system that yields accurate real-world results. *Hearing Review* online edition, www.hearingreview.com, October 2007.

Ross, M. (1980). Binaural versus monaural hearing aid amplification for hearing impaired individuals. In: Libby, E.R., Ed. *Binaural Hearing and Amplification.* Chicago: Zenetron, 1–21.

Walden, T.C. & Walden, B.E. (2005). Monaural versus binaural amplification for adults with impaired hearing. *Journal of the American Academy of Audiology,* 16, 574–584.

Can Preference for One or Two Hearing Aids Be Predicted?

Noble, W. (2006). Bilateral hearing aids: A review of self-reports of benefit in comparison with unilateral fitting. *International Journal of Audiology*, 45(Supplement 1), S63–S71.

The potential benefits of bilateral and unilateral hearing aids have been debated for years. Laboratory studies and clinical recommendations generally support the use of two hearing aids for individuals with bilateral hearing loss. Yet some field studies have produced equivocal reports. In his 2006 survey of bilateral and unilateral clinical field trials, William Noble discusses variables contributing to the lack of consensus and addresses a couple of commonly cited clinical rationales for bilateral hearing aid use. Though subject population, experimental design, degree of hearing loss, and usage patterns vary from one study to another, factors emerge that help determine likelihood of success in unilateral and bilateral hearing aid fitting.

The strong predisposition for clinicians to recommend bilateral hearing aid use may be based both on laboratory findings as well as common sense. Several studies have reported advantages of binaural listening (Dillon, 2001; McArdle et al., 2012), but clinicians often support their recommendation of two hearing aids with an analogy to binocular vision. Individuals with impaired vision no longer wear monocles (*with apologies to English detectives*) but instead opt for binocular corrective lenses. Noble argues that the visual analogy is not apt, partly because typical vision loss is not comparable to the typical hearing loss. Rather, vision loss that is treated by corrective lenses is most similar to a mild conductive hearing loss that is rarely treated with hearing aids. Cochlear receptor damage in sensorineural hearing loss, the most common type of hearing loss treated with hearing aids, introduces processing complexities that may not be adequately corrected by hearing aids that cannot address the auditory deficits.

Though Noble's comments are correct, the visual analogy is presented to hearing aid patients in an effort to explain how two hearing aids allow more effective use of bilateral listening cues, much as two corrective lenses can aid binocular, stereoscopic cues for depth and three-dimensional perception. Some patients may have difficulty understanding how two hearing aids can provide beneficial cues, especially in noise, instead of additional distraction. The visual analogy, while admittedly not perfect, is a way of explaining more simply the benefit of bilateral perceptual cues.

The other commonly cited clinical rationale for bilateral hearing aid use is related to the auditory deprivation effect (Silman et al., 1984). In individuals with bilateral hearing loss, there is concern that if only one hearing aid is used, the unaided ear will be deprived of sound and suffer additional deterioration. This appears to be a long-term change in the unaided ear, affecting word recognition scores but not pure-tone or speech-reception thresholds. Whether or not the unaided ear effect has implications for everyday hearing aid use and the ability to function in social and work-related situations requires further investigation to determine whether it is an important consideration for clinical recommendations.

Though most laboratory studies support the benefits of binaural listening, field studies and self-reports on bilateral hearing aid use have not always provided similar outcomes (Arlinger et al., 2003; Cox, 2011). For this reason, Noble reported on evidence from clinical trials to determine the conditions under which bilateral hearing aid use is most likely to be beneficial and to determine what patient attributes most support the recommendation of bilateral hearing aid use. Three of the reviewed studies were retrospective or reports from clinical patients occurring months or years after being fitted with unilateral or bilateral hearing aids. Two studies (Dirks & Carhart, 1962; Kochkin & Kuk, 1997) suggested that people who preferred bilateral hearing aid fittings had greater levels of hearing loss, though degree of hearing loss was not controlled. A third study by Noble, et al. (1995) carefully matched seventeen sets of unilateral and bilateral users according to degree of hearing loss. They examined speech reception and directional and distance spatial perception in aided and unaided conditions. Significant benefits were seen when comparing aided and unaided conditions but no differences were observed between unilaterally aided and bilaterally aided groups. The subject sample had mild-to-moderate hearing losses, so some caution should be taken when extrapolating these findings to individuals with more severe hearing losses.

In contrast, a study of new hearing aid users resulted in a two-to-one preference for unilateral use after a six-month period (Schreurs & Olsen, 1985). Individuals in this study wore one aid and two aids alternately for one week at a time, which arguably could have adversely influenced their acclimatization. Hearing aid users sometimes experience an extended period of adjustment to amplification (Keidser, 2009), which can affect their subjective judgments of sound quality and overall benefit (Bentler et al., 1993). For instance, occlusion and unnatural perception of one's own voice are qualities that can be annoying to new hearing aid users and are often more pronounced with bilateral hearing aids. Though these qualities almost always improve significantly with consistent bilateral use, it is not surprising that inexperienced, intermittent bilateral users might prefer the subjective sound quality of wearing one hearing aid at a time. Additionally, Schreurs and Olsen's study was conducted in 1985, at which time directional microphones were not in widespread use. Hearing aid users at that time often removed one hearing aid in noisy situations because bilateral omnidirectional microphones made surrounding noise sources too disruptive. A field study of unilateral versus bilateral use with modern hearing aids, allowing for adequate acclimatization, might yield different results.

Two follow-up studies examined subjects that were slightly younger than a typical clinical population (Brooks & Bulmer, 1981; Erdman & Sedge, 1981). Both of these studies found that a majority of subjects preferred the use of two hearing aids. These reports suggest that individuals whose activities require effective communication in challenging listening situations may prefer bilateral hearing aids. The remaining studies in Noble's review were crossover studies, or experiments in which clinical patients were randomly assigned to a unilateral or bilateral condition and crossed over to the other condition after several weeks. Stephens et al. (1991) found a greater degree of hearing loss and self-rated disability in the individuals who opted for bilateral hearing aid use; consistent with the retrospective reports of Dirks and Carhart (1962) and Kochkin and Kuk (1997) discussed earlier.

The studies examined in this literature review reveal several patterns. First, individuals with more severe hearing loss or perceived disability were more likely to prefer bilateral hearing aids. Subjects who preferred unilateral hearing aid use tended to have mild-to-moderate losses. Second, participants employed in dynamic listening situations preferred bilateral hearing aid use, suggesting that individuals whose regular activities require effective communication in a variety of contexts may be more likely to benefit from the use of two hearing aids (Noble & Gatehouse, 2006).

Noble points out that laboratory studies cannot adequately consider the range of experiences encountered by hearing aid users in everyday situations. Laboratory research isolates variables for study, with subjects responding to specific stimuli in isolated, carefully contrasted conditions, whereas in everyday life, hearing aid users encounter a wide range of listening situations ranging from single speech sources in quiet conditions to multiple speech sources in the presence of competing noise. Conversely, clinical field trials probe the subjective responses of hearing aid users in a variety of real-world situations, but it can be difficult to extricate the specific variables affecting their perceptions. Though their outcomes may not always appear to be in agreement, both types of study provide useful information to guide clinical practice.

It is clear that bilateral hearing aid use has potential to reduce listening effort (Feuerstein, 1992), improve speech understanding, localization, and receptiveness to lateral sounds (Noble & Gatehouse, 2006). Still, field studies consistently report a subset of patients that preference for unilateral hearing aid use. Whether environmentally or psychoacoustically motivated, the factors that underlie these preferences remain unclear. With consideration to documented benefits, bilateral hearing loss should first be treated with the prescription of bilateral hearing aids. Consideration for unilateral use should happen after the patient has adequate field experience and expresses subjective preference for the option of unilateral amplification.

References

Arlinger, S., Brorsson, B., Lagerbring, C., Leijon, A., Rosenhall, U. & Schersten, T. (2003). *Hearing Aids for Adults—benefits and costs*. Stockholm: Swedish Council on Technology Assessment in Health Care.

Bentler, R.A., Niebuhr, D.P., Getta, J.P. & Anderson, C.V. 1993b. Longitudinal study of hearing aid effectiveness. II. Subjective measures. *Journal of Speech and Hearing Research*, 36, 820–831.

Byrne, D., Noble, W. & LePage, B. (1992). Effects of long0term bilateral and unilateral fitting of different hearing aid types on the ability to locate sounds. *Journal of the American Academy of Audiology*, 3, 369–382.

Cox, R.M., Schwartz, K.S., Noe, C.M. & Alexander, G.C. (2011). Preference for one or two hearing aids among adult patients. *Ear and Hearing*, 32(2), 181–197.

Dillon, H. (2001). Monaural and binaural considerations in hearing aid fitting. In: Dillon, H., ed. *Hearing Aids*. Turramurra, Australia: Boomerang Press, 370–403.

Dirks, D. & Carhart, R. (1962). A survey of reactions from users of binaural and monaural hearing aids. *Journal of Speech and Hearing Disorders*, 27(4), 311–322.

Feuerstein, J.F. (1992). Monaural versus binaural hearing: Ease of listening, word recognition and attentional effort. *Ear & Hearing*, 13(2), 80–86.

Keidser, G., O'Brien, A., Carter, L., McLelland, M. & Yeend, I. (2009). Variation in preferred gain with experience for hearing-aid users. *International Journal of Audiology*, 2008 (47), 621–635.

Kochkin, S. & Kuk, F. (1997). The binaural advantage: Evidence from subjective benefit and customer satisfaction data. *The Hearing Review*, 4.

McArdle, R., Killion, M., Mennite, M. & Chisolm, T. (2012). Are Two Ears Not Better Than One? *Journal of the American Academy of Audiology*, 23, 171–181.

Noble, W. (2006). Bilateral hearing aids: A review of self-reports of benefit in comparison with unilateral fitting. *International Journal of Audiology*, 45(Supplement 1), S63–S71.

Noble, W. & Gatehouse, S. (2006). Effects of bilateral versus unilateral hearing aid fitting on abilities measured by the Speech, Spatial and Qualities of Hearing Scale (SSQ). *International Journal of Audiology*, 45(2), 172–181.

Noble, W., TerHorst, K. & Byrne, D. (1995). Disabilities and handicaps associated with impaired auditory localization. *Journal of the American Academy of Audiology*, 6(2), 129–140.

Silman, S., Gelfand, S. & Silverman, C. (1984). Late-onset auditory deprivation: Effects of monaural versus binaural hearing aids. *Journal of the Acoustical Society of America*, 76, 1357–1362.

Summarizing the Benefits of Bilateral Hearing Aids

Mencher, G.T. & Davis, A. (2006). Bilateral of unilateral amplification: is there a difference? A brief tutorial. *International Journal of Audiology*, 45 (S1), S3–11.

The decision to fit binaural hearing loss with bilateral hearing aids is influenced by a number of factors. The recommendation of two hearing aids may be contraindicated for financial reasons or because of near-normal hearing or profound loss in one ear, but the consensus among clinicians is that bilateral amplification is preferable for individuals with aidable hearing loss in both ears. Mencher and Davis examine a variety of considerations that may affect bilateral benefit, including speech intelligibility in noise, localization and directionality, sound quality, tinnitus suppression, binaural integration, and auditory deprivation. Research in these areas is discussed with reference to clinical indications for hearing aid fitting.

The authors begin with a clarification of the terms *binaural* and *bilateral*. They explain that a bilateral fitting refers to the use of hearing aids on both ears, whereas binaural hearing refers to the integration of signals arriving at two ears independently. They point out that standardization of these terms should help avoid confusion in the discussion of bilateral versus unilateral hearing aid fittings.

Because speech is the most important mode of everyday communication, studies of hearing aid benefit typically employ speech intelligibility measures in quiet and noisy conditions. Early studies investigating aided speech intelligibility yielded conflicting reports, with some in favor of bilateral amplification (Markle & Aber, 1958; Wright & Carhart, 1960; Olsen & Carhart, 1967; Markides, 1980) and others showing no difference between unilateral and bilateral fittings (Hedgecock & Sheets, 1958; DeCarlo & Brown, 1960; Jerger & Dirks, 1961).

Many early studies were criticized for methodological choices that could have obscured bilateral benefits, such as the use of a single noise source or test materials that were not representative of everyday, conversational speech. More recent work has examined speech intelligibility under conditions that more closely approximate real-world conditions, with multiple noise sources and sentence-based test materials. For instance, Kobler and Rosenhall (2002) studied intelligibility and localization for randomly presented speech from locations surrounding the listener, in the presence of speech-weighted noise from multiple loudspeakers. They found that bilateral amplification improved performance over unilateral fittings and unaided conditions. Their findings confirmed earlier work by Kobler, Rosenhall, and Hansson (2001) in which bilateral benefits were reported for speech recognition, localization, and sound quality, as well as more investigations such as that of McArdle and colleagues (2012).

Sound localization has implications for everyday environmental awareness as well as speech perception. Studies of auditory scene analysis (Bregman, 1990) underscore the importance of localization for identifying and attending to specific sound sources. This has specific relevance for understanding conversation in complex environments in which the speech must first be identified and separated from competing sound sources before higher-level processing can occur (Stevens, 1996). Therefore, the effect of hearing aids on localization is likely to impact an individual's overall ability to understand conversational speech in a noisy environment.

The physical presence of a hearing aid and earmold obscures some pinna-based localization cues; the use of bilateral hearing aids should aid horizontal localization under some circumstances. Individuals with moderate-to-severe hearing loss who wear only one hearing aid may hear some sounds only on the aided side, whereas binaural time and intensity cues may be preserved in a bilateral fitting. Current hearing aids with bilateral data exchange that can account for interaural phase and time cues may offer additional binaural localization benefits. Individuals in social situations are likely to be conversing with one or more people at roughly the same vertical elevation. Therefore, preservation of horizontal localization cues with bilateral hearing aids may outweigh the loss of pinna cues and may have more relevance for speech intelligibility, especially in noisy conditions.

Sound quality encompasses a number of attributes that include clarity, fullness, loudness, and naturalness. Bilateral hearing aid use may improve the quality of these attributes. Balfour and Hawkins (1992) examined eight sound-quality judgments for listeners with mild-to-moderate hearing loss, tested with unilateral and bilateral hearing aids. Subjects judged the sound quality of speech in quiet and noise, and music in a test booth, living room, and concert hall. Subjects had a significant preference for bilateral hearing aids for all sound quality dimensions, with clarity being ranked as the most important. This finding is in agreement with Erdman and Sedge (1981), who reported that clarity was the most significant benefit of bilateral amplification. Naidoo and Hawkins (1997) reported bilateral benefits for sound quality and speech intelligibility in high levels of background noise.

Tinnitus suppression is another area in which bilateral hearing aid use appears to offer an advantage over unilateral fittings. A questionnaire by Brooks and Bulmer (1981) found that 66.52% of bilaterally aided respondents experienced reduction in tinnitus versus only 12.7% with unilateral aids. Surr, Montgomery, and Mueller (1985) reported that about half of their subjects with tinnitus experienced partial or total relief from tinnitus with hearing aid use. Melin et al. (1987) found that there were differences in tinnitus relief based on the number of hours of use per day. Taken together, these studies suggest that individuals who suffer from tinnitus may experience from relief with hearing aids and are more likely to do so with consistent, bilateral hearing aid use.

There is evidence to suggest that some individuals experience better results with unilateral fittings. Binaural integration of simultaneous auditory signals in asymmetric hearing loss may have negative implications for bilateral hearing aid use. This was first described by Arkebauer, Mencher, and McCall (1971) who reported that amplified signals presented to two asymmetrically impaired ears resulted speech discrimination that was worse than the better ear alone and similar to the poorer ear alone. Hood and Prasher (1990) simulated bilateral hearing loss and found poorer speech discrimination ability when dissimilar distortion patterns were sent to each ear and significant improvement when identical distortion patterns were sent to the two ears. The results were interpreted to suggest that an inability to process incongruent or dissimilar speech input from both ears could contribute to the rejection of two hearing aids. Jerger et al. (1993) reported similar findings and explained that stimulation of the poor ear was interfering with the response of the better ear. He posited that binaural interference could affect approximately 10% of elderly hearing aid users. Binaural interference may be caused by age-related atrophy or corpus collosum demyelination resulting in poor inter-hemispheric transfer of auditory information (Chmiel et al. 1997) and individuals experiencing binaural interference may be likely to perform better with one hearing aid.

Auditory deprivation is commonly cited when recommending bilateral hearing aids. First described in 1984 by Silman, Gelfand, and Silverman, it was noted that in unilateral fittings on bilaterally impaired individuals, speech discrimination in the unaided ear was reduced relative to the aided ear. Pure-tone thresholds and speech reception thresholds were not affected. Gelfand, Silman, and Ross (1987) again found reduced speech-recognition scores over time (four to seventeen years) for unilaterally aided individuals. Hurley (1999) also reported that unilaterally aided subjects were more likely to experience monaural reductions in word recognition scores compared to bilaterally aided subjects. Subsequent studies determined that the auditory deprivation effect was reversible and subjects who were later fit with a second aid experienced improved word recognition (Silverman & Silman, 1990; Silverman & Emmer, 1993; Silman et al., 1992). Byrne and Dirks (1996) expanded on the concept of auditory deprivation, reporting that it may also affect localization and intensity discrimination. Though more research in this area may be warranted, Mencher and Davis note that the best way to treat auditory deprivation is to avoid it, with bilateral amplification as part of the solution.

Though research provides insight into possible predictors of success, an important measure of success can be found in post-fitting reports of unilateral and bilateral hearing aid users. Self-assessments of hearing handicap and disability can help hearing aid users express how their hearing aids affect important activities and everyday communication. Some investigations using self-assessment techniques have revealed a preference for bilateral hearing aid use (Chung & Stephens, 1983, 1986; Stephens et al., 1991) whereas others have revealed a preference for unilateral fittings (Cox, 2011). Because many factors contribute to an individual's preferences and perceived success, self-assessments should be used in combination with verification measures and consideration of individual attributes, such as age, experience with hearing aids, audiometric configuration, and speech discrimination ability. Most patient questionnaires target specific topics such as satisfaction, comfort, usage patterns, and speech intelligibility, so it may be useful to combine measures to gain comprehensive information about a patient's experience. The Speech, Spatial and Qualities of Hearing Scale (SSQ; Gatehouse & Noble, 2004), for instance, measures hearing disability in a variety of circumstances. Because it also examines directional, distance, and spatial perception, the SSQ may provide more insight into the effect of bilateral versus unilateral amplification on hearing in everyday situations.

Mencher and Davis's review suggests that there are numerous likely benefits to bilateral amplification for most, but not necessarily all, individuals. Bilateral hearing aids may offer an improved signal-to-noise ratio, reduced annoyance from tinnitus, improved sound quality and better localization in complex listening environments. Hearing aids are typically dispensed with a trial period of thirty to sixty days with relatively low financial risk to the patient in the event of a return. Therefore, it seems sensible to recommend bilateral fittings for candidates with bilateral hearing loss, knowing that a return of one hearing aid will be possible if contraindications to bilateral hearing aid use should arise during the trial period. Close monitoring of performance and comfort during the trial is essential, especially for individuals with asymmetrical hearing loss or a history of unilateral hearing aid use. In these cases, it may be necessary to reduce gain and output or increase compression in the poorer or previously unaided ear, to accommodate the likely inter-aural differences in acclimatization rate. Ears with more hearing loss and/or less aided experience generally take more time to adapt to amplification, so gradual adjustment and focused counseling may be necessary to eventually achieve satisfactory binaural balance.

The authors conclude by pointing out that the only way to know if a patient is successful with their hearing aids is to ask them! The interaction of several factors such as sound quality, localization, noise tolerance, loudness discomfort, and physical comfort will contribute to patient satisfaction. Ultimately, clinicians should develop a clinical strategy that employs objective and subjective measures to truly document benefit and satisfaction with the hearing aid fitting—be it unilateral or bilateral.

References

Arkebauer, H.J., Mencher, G.T. & McCall, C. (1971). Modification of speech discrimination in patients with binaural asymmetrical hearing loss. *Journal of Speech and Hearing Disorders*, 36, 208–212.

Balfour, P.B. & Hawkins, D.B. (1992). A comparison of sound quality judgments for monaural and binaural hearing aid processed stimuli. *Ear and Hearing*, 13, 331–339.

Bregman, A.S. (1990). Auditory Scene Analysis: The Perceptual Organization of Sound. Cambridge, Mass.: Bradford Books, MIT Press.

Brooks, D.N. & Bulmer, D. (1981). Survey of Binaural Hearing Aid Users. *Ear and Hearing*, 2, 220–224.

Byrne, D., Noble, W. & Lepage, B. (1992). Effects of long-term bilateral and unilateral fitting of different hearing aid types on the ability to locate sounds. *Journal of the American Academy of Audiology*, 3, 369–382.

Chmiel, R., Jerger, J., Murphy, E., Pirozzolo, F. & Tooley-Young, C. (1997). Unsuccessful use of binaural amplification by an elderly person. *Journal of the American Academy of Audiology*, 8, 1–10.

Chung, S. & Stephens, S.D.G. (1983). Binaural hearing aid use and the hearing measurement scale. IRCS Medical Science, 11:721–722. In W. Noble, 1998. *Self-Assessment of Hearing and Related Functions*, (London: Whurr).

Chung, S. & Stephens, S.D.G. (1986). Factors influencing binaural hearing aid use. *British Journal of Audiology*, 20, 129–140.

Cox, R.M., Schwartz, K.S., Noe, C.M. & Alexander, G.C. (2011). Preference for one or two hearing aids among adult patients. *Ear and Hearing*, 32(2), 181–197

DeCarlo, L.M. & Brown, W.J. (1960). The effectiveness of binaural hearing for adults with hearing impairment. *Journal of Auditory Research*, 1, 35–76.

Erdman, S. & Sedge, R. (1981). Subjective comparisons of binaural versus monaural amplification. *Ear and Hearing,* 2, 225–229.

Gatehouse, S. & Noble, W. (2004). The Speech, Spatial and Qualities of Hearing Scale (SSQ).*International Journal of Audiology,* 43, 85–99.

Gelfand, S., Silman, S. & Ross, L. (1987). Long term effects of monaural, binaural and no amplification in subjects with bilateral hearing loss. *Scandinavian Audiology,* 16, 201–207.

Hebrank, J. & Wright, D. (1974). Spectral cues used in the localization of sound sources on the median plane. *Journal of the Acoustical Society of America,* 56, 1829–1834.

Hedgecock, L.D. & Sheets, B.V. (1958). A comparison of monaural and binaural hearing aids for listening to speech. *Archives of Otolaryngology,* 68, 624–629.

Hood, J.D. & Prasher, D.K. (1990). Effect of simulated bilateral cochlear distortion on speech discrimination in normal subjects. *Scandinavian Audiology,* 19, 37–41.

Hurley, R.M. (1999). Onset of auditory deprivation. *Journal of the American Academy of Audiology,* 10, 529–534.

Jerger, J. & Dirks, D. (1961). Binaural hearing aids: An enigma. *Journal of the Acoustical Society of America,* 33, 537–538.

Jerger, J., Silman, S., Lew, H.L. & Chmiel, R. (1993). Case studies in binaural interference: converging evidence from behavioral and electrophysiological measures. *Journal of the American Academy of Audiology,* 122–131.

Kobler, S., Rosenhall, U., & Hansson, H. (2001). Bilateral hearing aids - effects and consequences from a user perspective. *Scandinavian Audiology,* 30, 223–235.

Kubler, S. & Rosenhall, U. (2002). Horizontal localization and speech intelligibility with bilateral and unilateral hearing aid amplification. *International Journal of Audiology,* 41, 392–400.

Markides, A. (1980). Binaural Hearing Aids. NY: Academic Press.

Markle, D.M. & Aber, W. (1958). A clinical evaluation of monaural and binaural hearing aids.*Archives of Otolaryngology*, 67, 606–608.

McArdle, R., Killion, M., Mennite, M. & Chisolm, T. (2012). Are Two Ears Not Better Than One?*Journal of the American Academy of Audiology*, 23, 171–181.

Mehrgardt, S. & Mellert, V. (1977). Transformational characteristics of the external human ear.*Journal of the Acoustical Society of America* 61, 1567-1576.

Melin, L., Scott, B., Lindberg, P. & Lyttkens, L. (1987). Hearing aids and tinnitus - an experimental group study. *British Journal of Audiology*, 21, 91–97.

Mencher, G.T. & Davis, A. (2006). Bilateral of unilateral amplification: is there a difference? A brief tutorial. *International Journal of Audiology*, 45(S1), S3–11.

Middlebrooks, J.C. & Green, D.M. (1991). Sound localization by human listeners. *Annual Review of Psychology*, 42, 135–159.

Naidoo, S.V. & Hawkins, D.B. (1997). Monaural/binaural preferences: effect of hearing aid circuit on speech intelligibility and sound quality. *Journal of the American Academy of Audiology*, 8, 188–202.

Noble, W., Sinclair, S. & Byrne, D. (1998). Improvement in aided sound localization with open earmolds: Observations in people with high-frequency hearing loss. *Journal of the American Academy of Audiology*, 9, 25–34.

Olsen, W.R. & Carhart, R. (1967). Development of test procedures for evaluation of binaural hearing aids. *Bulletin of Prosthetics Research*, 10, 22–49.

Searle, C., Braida, L., Cuddy, D. & Davis, M. (1975). Binaural pinna disparity: Another auditory localization cue. *Journal of the Acoustical Society of America*, 57, 448–455.

Silverman, C & Emmer, M.B. (1993). Auditory deprivation and recovery in adults with asymmetric sensorineural hearing impairment. *Journal of the American Academy of Audiology*, 4, 338–346.

Silverman, C. & Silman, S. (1990). Apparent auditory deprivation from monaural amplification and recovery with binaural amplification: 2 case studies. *Journal of the American Academy of Audiology*, 1, 175–180.

Silman, S., Gelfand, S. & Silverman, C. (1984). Late-onset auditory deprivation: effects of monaural versus binaural hearing aids. *Journal of the Acoustical Society of America*, 76, 1357–1362.

Silman, S., Silverman, C.A., Emmer, M.B. & Gelfand, S. (1992). Adult-onset auditory deprivation. *Journal of the American Academy of Audiology*, 3, 390–396.

Stephens, S.D., Callaghan, D.E., Hogan, S., Meredith, R., Rayment, A. & Davis, A.C. (1991). Acceptability of binaural hearing aids: a crossover study. *Journal of the Royal Society of Medicine*, 84, 267–269.

Stevens, K. (1996). Amplitude-modulated and unmodulated time-varying sinusoidal sentences: the effects of semantic and syntactic context. *Doctoral dissertation, Northwestern University* (University of Michigan Press, AAT 9632785).

Surr, R.K., Montgomery, A.A. & Mueller, H.G. (1985). Effect of amplification on tinnitus among new hearing aid users. *Ear and Hearing*, 6, 71–75.

Wright, H.N. & Carhart, R. (1960). The efficiency of binaural listening among the hearing impaired. *Archives of Otolaryngology*, 72, 789–797.

What's New with NAL-NL2?

Keidser, G., Dillon, H., Flax, M., Ching, T. & Brewer, S. (2011). The NAL-NL2 prescription procedure. *Audiology Research*, 1 (e24), 88–90.

For years, the National Acoustics Laboratory's NAL-NL1 has been the benchmark for compressive, independently derived, prescriptive formulas (Dillon, 1999). The recently introduced NAL-NL2 advances their original formula with knowledge gained from a wealth of empirical data collected with NAL-NL1 (Keidser et al., 2011). Similarities between NAL-NL1 and NAL-NL2 include: primary goals of maximizing speech intelligibility while not exceeding overall normal loudness at a range of input levels and the use of predictive models for speech intelligibility and loudness (Moore & Glasberg, 1997; 2004).

The speech intelligibility model used in both NAL-NL1 and NAL-NL2 differs from the Speech Intelligibility Index (SII; ANSI, 1997). The ANSI SII assumes that, regardless of hearing loss, speech should be fully understood when all speech components are audible. Included in NAL-NL1 is a modification to the SII proposed by Ching and colleagues (1998). This modification or effective audibility factor assumes that as hearing loss becomes more severe less information can be extracted from the speech signal. More recent data have been collected to derive an updated effective audibility factor for use with NAL-NL2 (Keidser et al., 2011).

The NAL-NL2 formula includes constraints to prevent compression ratios from exceeding a maximum value for a given frequency or degree of hearing loss. Modifications were based on data suggesting that users with severe or profound hearing loss prefer lower compression ratios than those prescribed by NAL-NL1, when fitted with fast-acting compression (Keidser et al., 2007). However, there is evidence to suggest that higher compression ratios could be used in this population with slow-acting compression. Therefore, in the case of severe or profound hearing losses, the new formula prescribes lower compression ratios for fittings with fast-acting compression than those with slow-acting compression. For mild and moderate losses, compression speed does not affect prescribed compression ratios.

Based on experimental outcomes with NAL-NL1 fittings, the development of NAL-NL2 took various attributes of the hearing aid user into consideration, such as gender, binaural listening, experience level, age, and language. In the case of gender, Keidser and Dillon (2006) studied the real-ear insertion gain measurements for the preferred frequency responses of 187 adults, finding that regardless of experience or degree of hearing loss, female participants preferred an average of 2dB less gain than male participants. As a result, gender differences are factored into each fitting.

The NAL-NL2 method still prescribes greater gain for monaural fittings than it does for binaural fittings. This difference is similar to the NAL-NL1 formula (Dillon et al., 2010). Recent studies suggest that the binaural to monaural loudness difference may be less than previously indicated (Epstein & Florentine, 2009). For symmetrical hearing losses, the binaural difference ranges from 2dB for inputs below 50dB SPL to 6dB for inputs above 90dB SPL, so binaurally fitted users will have higher prescribed compression ratios than monaural users. For asymmetrical losses, the binaural correction decreases as the asymmetry increases.

Experience with hearing aids as it relates to degree of hearing loss is a consideration in the NAL-NL2 formula. Keidser and her colleagues (2008) found that with increasing severity of hearing loss new users prefer progressively less prescribed gain than experienced hearing aid users. Although this observation does not agree with several other studies (e.g., Convery et al., 2005; Smeds et al., 2006), the NAL-NL2 recommends gain adaptation for new hearing aid users with moderate or severe hearing loss. Further details of this discrepancy will be addressed in future publication (Keidser 2012, personal communication).

The developers of the NAL-NL2 formula determined that adults with mild-to-moderate hearing loss preferred less overall gain for 65dB inputs than would be prescribed by NAL-NL1 (Keidser et al., 2008). This is corroborated by other studies (Smeds et al., 2006; Zakis et al., 2007) in which hearing aid users with mild-to-moderate hearing loss preferred less gain for high- and low-level inputs. These reports indicate that participants generally preferred slightly less gain and higher compression ratios than those prescribed by NAL-NL1, a preference that was incorporated into the revised prescriptive procedure.

The NAL-NL2 also takes the hearing aid user's language into consideration. For speakers of tonal languages, slightly more low-frequency gain is prescribed. Increased gain in the low-frequency region more effectively conveys fundamental frequency information, an especially important cue for recognition of tonal languages.

Like its predecessor, the NAL-NL2 fitting formula leverages theoretical models of intelligibility and loudness perception to maximize speech recognition without exceeding normal loudness. The revised formula takes into consideration a number of factors other than audiometric information and benefits from extensive empirical data collected using NAL-NL1. Ultimately, the NAL-NL2 procedure results in a slightly flatter frequency response, with relatively more gain across low and high frequencies and less gain in the mid-frequency range than in the NAL-NL1 formula. The study of objective performance and subjective preference with hearing aids is constantly evolving and the NAL-NL2 prescriptive method may be a step toward achieving increased acceptance by existing hearing aid users and improved spontaneous acceptance by new hearing aid users.

References

American National Standards Institute (1997). Methods for calculation of the speech intelligibility index. *ANSI S3.5-1997*. Acoustical Society of America, New York.

Ching, T., Dillon, H. & Byrne, D. (1998). Speech recognition of hearing-impaired listeners: Predictions for audibility and the limited role of high frequency amplification. *Journal of the Acoustical Society of America, 103*(2), 1128–1140.

Convery, E., Keidser, G., & Dillon, H. (2005). A review and analysis: Does amplification experience have an effect on preferred gain over time? *Australian and New Zealand Journal of Audiology, 27*(1), 18–32.

Dillon, H. (1999). Page Ten: NAL-NL1: A new procedure for fitting non-linear hearing aids. *Hearing Journal, 52*, 10–16.

Dillon, H., Keidser, G., Ching, T. & Flax, (2010). The NAL-NL2 Prescription Formula, conference paper, *Audiology Australia 19th National Conference*, May 2010.

Epstein, M. & Florentine, M. (2009). Binaural loudness summation for speech and tones presented via earphones and loudspeakers. *Ear and Hearing*, 30(2), 234–237.

Keidser, G. & Dillon, H. (2006). What's new in prescriptive fittings down under? In: Palmer, C.V., Seewald, R. (Eds.), Hearing Care for Adults 2006. Phonak AG, Stafa, Switzerland, 133–142.

Keidser, G., Dillon, H., Dyrlund, O., Carter, L. & Hartley, D. (2007). Preferred low and high frequency compression ratios among hearing aid users with moderately severe to profound hearing loss. *Journal of the American Academy of Audiology*, 18(1), 17–33.

Keidser, G., O'Brien, A., Carter, L., McLelland, M. & Yeend, I. (2008). Variation in preferred gain with experience for hearing aid users. *International Journal of Audiology*, 47(10), 621–635.

Keidser, G., Dillon, H., Flax, M., Ching, T. & Brewer, S. (2011). The NAL-NL2 prescription procedure. *Audiology Research*, 1(e24), 88–90.

Moore, B.C.J. & Glasberg, B. (1997). A model of loudness perception applied to cochlear hearing loss. *Audiotory Neuroscience, 3*, 289–311.

Moore, B.C.J. & Glasberg, B. (2004). A revised model of loudness perception applied to cochlear hearing loss. *Hearing Research*, 188, 70–88.

Smeds, K., Keidser, G., Zakis, J., Dillon, H., Leijon, A., Grant, F., Convery, E. & Brew, C. (2006). Preferred overall loudness. II: Listening through hearing aids in field and laboratory tests. *International Journal of Audiology*, 45(1), 12–25.

Zakis, J., Dillon, H. & McDermott, H.J. (2007). The design and evaluation of a hearing aid with trainable amplification parameters. *Ear and Hearing*, 28(6), 812–830.

The DSL v5.0a Is a Successful Fitting Formula for Adults

Polonenko, M.J., Scollie, S.D., Moodie, S., Seewald, R.C., Laurnagaray, D., Shantz, J., & Richards, A. (2010). Fit to targets, preferred listening levels and self-reported outcomes for the DSL v5.0a hearing aid prescription for adults. *International Journal of Audiology*, 49, 550–560.

The importance of perceived benefit for successful hearing aid fittings is well established. According to two MarkeTrak studies by Sergei Kochkin (2005, 2007), perceived benefit was the number one factor contributing to hearing aid user satisfaction. Similarly, the lack of benefit was the most commonly cited reason for hearing aid returns. Perceived benefit from hearing aids may be determined by a number of factors, but the appropriateness of the individually fitted gain is one of the main contributors (Cox & Alexander, 1994).

The Desired Sensation Level (DSL) prescriptive method was originally developed for children and prescribes targets that are generally very close to children's preferred listening levels. However, DSL v4.1 targets have been found to prescribe gain that is 9 to 11 dB greater than adult preferred listening levels (Scollie et al., 2005). Therefore, DSL v5.0a was developed with lower perceived loudness levels, ones that more closely approximate the needs of adult hearing aid users.

The success of a hearing aid prescription can be measured in terms of clinical efficacy, or how closely the hearing aid settings achieve a desired clinical result or test outcome. One such measure is the Preferred Listening Level (PLL). The PLL is defined as "the sound pressure level at the eardrum that the person chooses or prefers for listening to hearing aid amplified speech"(Cox & Alexander, 1994) and represents a compromise between comfort, intelligibility, background noise, and distortion (Cox, 1982). One method of measuring the PPL is by instructing listeners to adjust the volume setting of their hearing instruments to the level that sounds best to them as they listen to speech presented at a conversational level.

A related but different way to determine the success of a hearing aid fitting strategy is measure effectiveness, or how well hearing aid settings help the user function in real-world situations. One commonly used measure of hearing aid effectiveness is the Client Oriented Scale of Improvement (COSI) (Dillon et al., 1997). On the COSI questionnaire, the hearing aid user lists up to five typical listening situations in which he or she struggles to hear or would like to hear better. Following a period of acclimatization, they rate the degree of perceived change in these situations as well as their final ability to function in each situation.

Although the DSL v5.0a prescriptive method was specifically developed for adults with acquired hearing loss, there have been relatively few studies evaluating it. Therefore the current authors sought to determine the electroacoustic feasibility, clinical efficacy, and effectiveness with adult hearing aid users. They had three primary goals:

1. To measure final fit versus targets in a clinical environment

2. To evaluate the preferred listening levels (PLLs) of adults versus the DSL v5.0a targets

3. To measure the effectiveness of the DSL v5.0a prescription as reported on the COSI

Thirty subjects with predominantly sensorineural hearing loss participated in the study. Nineteen were new hearing aid users and eleven were experienced hearing aid users. Twenty-four were fitted binaurally and six were monaural users. Subjects were fitted in private clinics and the audiologists were specifically instructed to program and adjust the instruments to meet the patients' needs, rather than to meet prescriptive targets.

Hearing aid fittings were matched to DSL 5.0 prescribed targets and verified with simulated real ear measurements, to ensure consistency between test sites and to promote replicable measures. Hearing aids were set to their primary programs and were measured in 2cc couplers, after individual Real Ear to Coupler Differences (RECD) were measured. Following electroacoustic measures, the aids were fitted to the patients' ears and adjustments were made based on patients' subjective satisfaction. These procedures were not carried out according to any protocol established by the authors; the audiologists conducted fine-tuning adjustments as needed for each individual. After an approximately thirty-day period, subjects returned to the clinics for fine tuning. After a total acclimatization period of ninety days, preferred listening levels (PLLs) and COSI outcome evaluations were conducted.

Electroacoustic analyses revealed that the clinical fittings were significantly correlated with the DSL v5.0a targets. Sixty-eight percent of initial fittings were within 2.9 to 4.2 dB of target and 95% were within 5.8 to 8.4 dB of target across frequencies. These results contrast with previous research using NAL-R and NAL-NL1 targets, in which initial fittings differed from targets by 10–15dB. (Sammeth et al., 1993; Aazh & Moore, 2007).

Preferred listening levels (PLLs) were compared to targets and initial fittings and differed by only about 2dB. The DSL v5.0a targets were on average 2.6dB lower than PLLs and 1.95dB lower than initial fittings. Furthermore, DSL v5.0a targets were significantly correlated with PLLs at all frequencies and the targets and PLLs did not differ significantly as a function of degree of hearing loss. The authors noted a trend for higher PLLs than targets at 250Hz, indicating that some users preferred more low-frequency output than prescribed.

COSI ratings of real-world performance were obtained at the ninety-day appointment. The top five situations in which subjects hoped to hear better were similar to those chosen by subjects in the COSI normative study (Dillon et al., 1999). They include,

1. Conversation with a group in noise
2. Conversation with a group in quiet
3. Conversation with one or two partners in noise
4. Listening to the television or radio
5. Conversation with one or two partners in quiet

Subjects were asked to rate the degree of change in their hearing with amplification as well as the final hearing ability (or hearing aid performance) in these situations. Results indicated that they judged their hearing to be "better" or "much better" for 83% of the fittings, which compares well to the normative results obtained by Dillon et al. (1999) of 80%. For final hearing ability, 93% of the current respondents reported hearing 75% of the time (a COSI rating of four or better) as compared to 90% of the normative study participants.

The purpose of the current study was to determine if DSLv5.0a prescriptive targets, developed for adults, provided electroacoustically appropriate fittings and subjectively favorable real-world results. Indeed, clinician-adjusted fittings were within 10dB of prescriptive targets for 92% of the subjects. Targets also closely approximated preferred listening levels, which is particularly important because prior studies showed DSL v4.1 targets were generally higher than adults' preferred levels. COSI measurements indicated positive ratings for benefit and communication performance that were similar or slightly better than those obtained for the normative study population.

An incidental finding of the current study was that instruments with more than six channels of processing may meet prescriptive targets more accurately than those with only six channels. This was not specifically studied in the current paper, but the authors provided a matrix of number of channels versus errors in matching to target, showing that instruments with more than six channels yielded fewer and smaller errors than those with only six channels of processing. This result is probably consistent with clinical observations, in which sophisticated hearing aid circuits with more channels of processing often provide better fittings than instruments with fewer channels. The importance of this factor may depend on the client's hearing loss. Gently sloping audiometric configurations may generally require fewer channels to meet targets.

The results of this study show that in a group of adults preferred listening levels and positive real-world outcomes were achieved with programs matched to DSL v5.0a targets, at least in quiet situations. In noisy listening situations, participants may have accessed alternate memories with directionality and noise reduction, causing amplification characteristics to differ from DSL settings. Even if this is the case, the current study shows that the DSL v5.0a prescriptive measure for adults yields a close approximation to patient preferred settings for a wide range of hearing losses.

References

Aazh, H. & Moore, B.C.J. (2007). The value of routine real ear measurement of the gain of digital hearing aids. *Journal of the American Academy of Audiology*, 18, 653–664.

Cox, R.M. (1982). Functional correlates of electroacoustic performance data. In G.A. Studebaker & F.H. Bess (Eds.) *The Vanderbilt Hearing Aid Report*. (78–84). Parkton, MD: York Press,

Cox, R.M. & Alexander, G.C. (1994). Prediction of hearing aid benefit: the role of preferred listening levels. *Ear and Hearing*, 15(1), 22–29.

Dillon, H., James, A., & Ginis, J. (1997). Client Oriented Scale of Improvement (COSI) and its relationship to several other measures of benefit and satisfaction provided by hearing aids. *Journal of the American Academy of Audiolog*, 8, 27–43.

Dillon, H., Birtles, G., & Lovegrove, R. (1999). Measuring the outcomes of a National Rehabilitation Program: normative data for the Client Oriented Scale of Improvement (COSI) and the Hearing Aid User's Questionnaire (HAUQ). *Journal of the American Academy of Audiology*, 10, 67–79.

Kochkin, S. (2005). MarkeTrak VII: Customer satisfaction with hearing instruments in the digital age. *Hearing Journal*, 58(9), 30–43.

Kochkin, S. (2008). MarkeTrak VII: Obstacles to adult non-user adoption of hearing aids. *Hearing Journal*, 60(4), 24–51.

Polonenko, M.J., Scollie, S.D., Moodie, S., Seewald, R.C., Laurnagaray, D., Shantz, J., & Richards, A. (2010) Fit to targets, preferred listening levels and self-reported outcomes for the DSL v5.0a hearing aid prescription for adults. *International Journal of Audiology*, 49, 550–560.

Sammeth, C., Peek, B., Bratt, G., Bess, F., & Amberg, S. (1993). Ability to achieve gain/frequency response and SSPL-90 under three prescription formulas with in-the-ear hearing aids. *Journal of the American Academy of Audiology*, 4, 33–41.

Scollie, S., Seewald, R., Cornelisse, L., Moodie, S., Bagatto, M., et al. (2005). The Desired Sensation Level Multistage Input/Output Algorithm. *Trends in Amplification*, 4(9), 159–197.